A Practical Guide for SystemVerilog Assertions

A Practical Guide for SystemVerilog Assertions

by

Srikanth Vijayaraghavan

Meyyappan Ramanathan

Springer

Srikanth Vijayaraghavan & Meyyappan Ramanthan
Synopsys, Inc.
Mountan View, CA
USA

A Practical Guide for SystemVerilog Assertions

Library of Congress Control Number: 2005049012

ISBN 0-387-26049-8 e-ISBN 0-387-26173-7
ISBN 9780387260495

Printed on acid-free paper.

Printed in the United States of America.

9 8 7 6 5 4 3 2

springer.com

Dedication

To my wonderful wife, Anupama – I could not have done this without your love and support.

Srikanth Vijayaraghavan
Synopsys, California

To my wife, Devi, and my children, Parvathi and Aravind – thank you for your patience during the long hours spent in completing the book.

Meyyappan Ramanathan
Synopsys, California

Table of Contents

List of Figures

Chapter 0: Assertion Based Verification

Chapter 1: Introduction to SVA

Chapter 2: SVA Simulation Methodology

Chapter 3: SVA for Finite State Machines

Chapter 4: SVA for Data Intensive Designs

Chapter 5: SVA for Memories

Chapter 7: Checking the Checker

List of Tables

Chapter 6: SVA for Protocol Interface

Chapter 7: Checking the Checker

Foreword

by
Ira Chayut, Verification Architect
Nvidia Corporation

When Gateway Design Automation, Inc. created Verilog in the mid-1980's, the process of integrated circuit design was very different than it is today. The role of Verilog, as well as its capability, has evolved since its inception into today's SystemVerilog.

The task of ASIC Functional Verification is becoming increasingly difficult. How difficult is a matter of conjecture and argument. In 2001, Andreas Bechtolsheim, Cicso Systems engineering vice president, was quoted in *EE Times* with one of the higher estimates:

> Design verification still consumes 80 percent of the overall chip development time[1]

In contrast, an *EE Times* poll that was taken in 2004 of 662 professionals at the Design Automation Conference placed functional verification as 22 percent of the integrated design process[2].

The gap between 22 percent and 80 percent is indicative of how vague the delineation between verification and the other "stages" of integrated circuit design and development. Many "verification" efforts are implemented by the design engineers themselves, but are still part of the verification process and can benefit from the same tools that assist dedicated verification professionals.

Regardless of the actual percentage (assuming that it could be accurately measured), Functional Verification of an integrated circuit design is a significant fraction of the total effort. Verification is also a critical step to shippable first silicon. Even as the costs of the masks run over $1 million, that figure can be dwarfed by the lost-opportunity of the weeks it takes for each re-spin. Any tools that can reduce the cost of verification and increase the probability of shipping early silicon should be adopted aggressively.

[1] http://www.eedesign.com/article/printableArticle.jhtml?articleID=17407503
[2] http://www.eetimes.com/showArticle.jhtml?articleID=21700028

While assertions have been a part of software development for many years, Assertion-Based Verification (ABV) has recently become popular. In some ways, this is odd, as the process of hardware specification has become more similar to software design. However, the properties that we wish to declare and assert in a hardware design are fundamentally different than those in the software world.

The difference between hardware and software programming models is ***time***. Hardware languages, such as Verilog, have mechanisms to represent the passage of time and procedural programming languages (C, C++, Java, etc.) do not. So, it is not surprising that the software methods of specifying assertions did not have a way of incorporating time.

SystemVerilog, the most recent descendent of Gateway's Verilog, includes SystemVerilog Assertions (SVA) – a set of tools to allow engineers to include ABV into their designs. SVA has a rich syntax to support time within sequences, properties, and (ultimately) assertions.

With SVA, design and verification engineers can encode the intended behavior of hardware designs and can create thorough checks for bus protocols. These (relatively) terse descriptions can be used in simulation, in formal verification, and as additional documentation for the design.

It is clear that SVA will have a major impact on how integrated circuits are designed and verified. To benefit from this impact, you need to learn the syntax of SVA and how to apply it to your own design. This book can help you learn and apply SVA. It uses examples, including the PCI bus protocol, to illustrate how to write SVA and their simulation results.

The detailed examples of the SVA language within this book are very helpful to understanding the concepts and syntax of time-based assertions. They make the book what it is and are essential in all SystemVerilog design and verification engineers' library.

As a final note, Stevie, my daughter, claims that no one ever reads the foreward of books. If you did take the time to read this, please let her know by sending her a brief e-mail at: steviechayut@gmail.com.

Thanks
Ira

Preface

It was the middle of the year 2002 and we received an email from our manager. It said, "Who would like to pick up the support for OVA?" Our first thoughts were "what the heck is OVA?" After talking to a few other engineers, we figured out that it was a subset of "open VERA language." OVA stands for "Open VERA Assertions" and it is a declarative language that can describe temporal conditions. As always, to satisfy our technical thirst, we agreed to pick up the support for OVA. We learned the language in a couple of months and started training customers, training around 200 customers in less than 6 months. The way customers were flooding the class rooms really impressed us. We were convinced that this is the next best thing in verification domain. While customers were getting trained in a hurry, they were not developing any OVA code. This was a new dimension of verification technique and the language was new. The tools were just starting to support these language constructs. There was not much intellectual property (IP) available. Naturally, customers were not as comfortable as we thought they should be.

In the meantime, Synopsys Inc. had donated the Open VERA language to the Accellera committee to be part of the SystemVerilog language. Several other companies made contributions for the formation of the new SystemVerilog language. The Accellera committee ratified the SystemVerilog 3.1 language as a standard at DAC 2004. The SystemVerilog language included the assertion language as part of the standard. This is commonly referred to as "SystemVerilog Assertions" (SVA). We continued in the path of training customers in Assertion based verification, only now we were teaching SVA. We could see clearly that customers were more comfortable with the pre-developed assertion libraries, but they were reluctant to write custom assertion code. What could be holding them back? Was it the tools? No, the tools were ready. Was it the language? Maybe, but it is a standard now, so that wasn't necessarily the case.

After a few lengthy discussions, we realized that the lack of examples to demonstrate SVA language constructs could be holding back customers from using this new technology. The lack of expertise typically contributes to slow adoption. This is when we thought an SVA cookbook might help—a book of examples, a book that could act as a tutorial, a book that could teach the language. And that is how this project started. We have made an effort to write what we learned from teaching this subject for the past two years.

While there is much more to learn in this area, this is just an effort to share what we have learned.

How to read this book.

This book is written in a way such that engineers can get up to speed with SystemVerilog assertions quickly.

Chapters 0, 1 and 2 are sufficient to learn the basics of the syntax and some of the common simulation techniques. After reading these three chapters, the user should be able to write assertions for their design/verification environment.

Chapter 3, 4, 5 and 6 are cookbooks for different types of designs. A user can refer to these chapters if they come across similar designs in their own environment and use these chapters as a starting point for writing assertions. These chapters can also be used as a tutorial.

If you are someone new to assertion based verification, you need to read chapters 0 through 2 before reading the other chapters. If you are familiar with SVA language, you can refer to these chapters on an as needed basis.

Chapter 0 - This is a white paper on "Assertion based verification (ABV)" methodology. It introduces the concept of ABV and the importance of function coverage.

Chapter 1 - Discusses SVA syntax with simple examples and goes through a detailed analysis of the execution of SVA constructs in dynamic simulation. Simulation waveforms and event tables are included for the reader's reference. To know the details of every SVA construct, the user should refer to the SystemVerilog 3.1 a LRM (Chapter 17).

Chapter 2 – Uses a system example to illustrate SVA simulation methodology. Topics cover protocol extraction, simulation control and functional coverage.

Chapter 3 - Illustrates how to verify FSMs with SVA, uses two different FSM models as examples.

Chapter 4 – Illustrates verification of a data path using SVA. A partial JPEG design is used to demonstrate verification of both control signals and data using SVA.

Chapter 5 – Illustrates verification of a memory controller using SVA. The controller supports different types of memories such as SDRAM, SRAM, Flash, etc.

Chapter 6 – Illustrates verification of a PCI local bus based system using SVA. A sample PCI system configuration is used and various PCI protocols are verified using SVA.

Chapter 7 – Illustrates a sample testbench for verifying the assertions. It also discusses the theory behind verifying the accuracy of an assertion.

A CD-ROM is included with the book. All the examples shown in the book can be run with VCS 2005.06 release. Sample scripts to run the examples are included. VCS is a registered trademark of Synopsys Inc.

Acknowledgements

The authors would like to convey their sincere thanks to the following individuals that have contributed immensely for the completion of this book.

Anupama Srinivasa, DSP Solutions Architect, AccelChip, Inc.
Jim Kjellsen, Staff Applications Consultant, Synopsys, Inc.
Juliet Runhaar, Senior Applications Consultant, Synopsys, Inc.

We would also like to thank the following individuals for reviewing the book and providing several constructive suggestions:

Ira Chayut, Bohran Roohipour, Irwan Sie, Ravindra Viswanath, Parag Bhatt, Derrick Lin, Anders Berglund, Steve Smith, Martin Michael, Jayne Scheckla, Rakesh Cheerla, Satish Iyengar

Useful Web links

www.systemVerilogforall.com – Page maintained by us that provides tips, examples and discussions on SystemVerilog language.

www.accellera.org – Official website of Accellera committee. The SystemVerilog LRM can be downloaded from this site. There are also several other useful papers and presentations on the latest standards.

Chapter 0

ASSERTION BASED VERIFICATION

Use of assertions justified

The growing complexity and size of digital designs have made functional verification a huge challenge. In the last decade several new technologies have emerged in the area of verification and some of them have captured their place as a requirement in the verification process.

Figure 0-1 shows a block diagram of a verification environment that is adopted by a vast majority of verification teams. There are two significant pieces of technology that are used by almost all verification engineers:

1. A constrained random testbench
2. Code coverage tool

The objective is to verify the design under test (DUT) thoroughly and make sure there are no functional bugs. While doing this, there should be a way of measuring the completeness of verification. Code coverage tools provide a first level measure on the verification completeness. The data collected during code coverage has no knowledge of the functionality of the design but provides information on the execution of the code line by line. By guaranteeing that every line of the DUT executed at least once during simulations, a certain level of confidence can be achieved and code coverage tolls can help achieve that. Last but not the least, the process of verification should be completed in a timely fashion. It is a well-known fact that the worst bottleneck for any verification environment is performance.

Traditionally, designs are tested with stimulus that verifies a specific functionality of the design. The complexity of the designs forces verification engineers to use a random testbench to create more realistic verification scenarios. High-level verification languages like OPEN VERA are used extensively in creating complex testbenches.

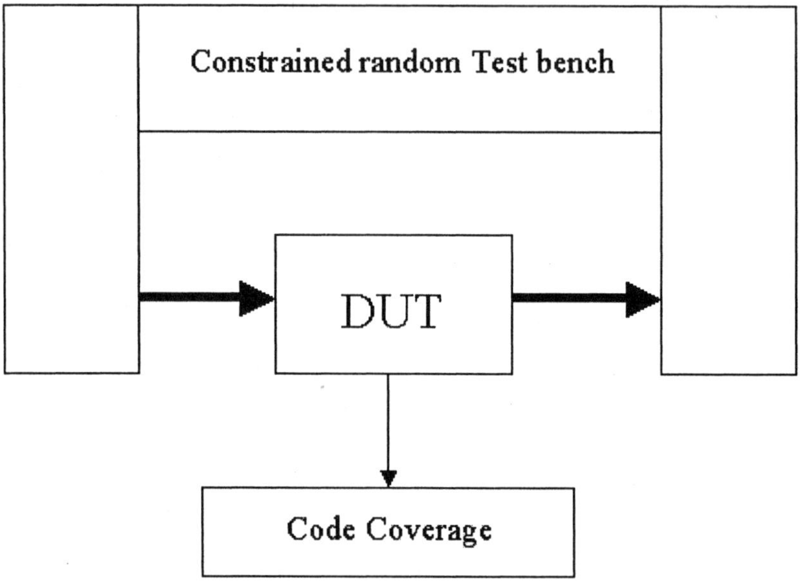

Figure 0-1. Before Assertion based verification

The testbenches normally perform three different tasks:

1. Stimulus generation.
2. Self-checking mechanisms.
3. Functional coverage measurement.

The first and foremost aim of a testbench is to create good quality stimulus. Advanced languages like OPEN VERA provide built-in mechanisms to create complex stimulus patterns with ease. These languages support object-oriented programming constructs that help improve the stimulus generation process and also the re-use of the testbench models.

A testbench should also provide excellent self-checking mechanisms. It is not always possible to debug the design in post-processing mode.

Mechanisms like waveform debugging are prone to human error and are also not very feasible with the complex designs of today. Every test should have a way of checking the expected results automatically and dynamically. This will make the debugging process easy and also make the regression tests more efficient. Self-checking processes usually targets two specific areas:

1. Protocol checking
2. Data checking

Protocol checking targets the control signals. The validity of the control signals is the heart of any design verification. Data checking deals with the integrity of the data being dealt with. For example, are the packets getting transferred without corruption in a networking design? Data-checking normally requires some level of formatting and massaging that is usually taken care of within the testbench environment effectively.

Functional coverage provides a measure of verification completeness. The measurement should contain information on two specific items:

1. Protocol coverage
2. Test plan coverage

Protocol coverage gives a measure on exercising the design for all valid and invalid design conditions. In other words, it is a measure against the functional specification of the design that confirms that all possible functionality has been tested. Test plan coverage, on the other hand, measures the exhaustiveness of the testbench. For example, did the testbench create all possible packet sizes, did the CPU write or read to all possible memory address spaces? Protocol coverage is measured directly from the design signals, and the test plan coverage can be easily measured with built-in methods within the testbench environment.

SystemVerilog assertions modify the verification environment in a manner such that the strengths of different entities are leveraged to the maximum. Figure 0-2 shows the modified block diagram for the verification environment that includes Assertion Based Verification (ABV).

There are two categories discussed in the different pieces of the testbench, which are addressed in detail by SystemVerilog assertions (SVA):

1. Protocol checking
2. Protocol coverage

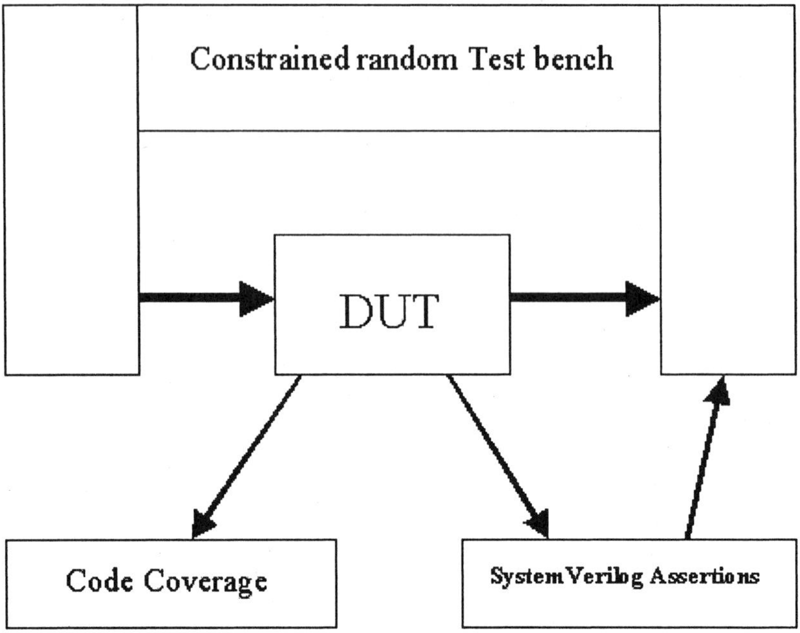

Figure 0-2. After SystemVerilog assertions

These two categories are closer to the design signals and can be managed more efficiently within SVA than by the testbench. By connecting these assertions directly to the design, the performance of the simulation environment increases tremendously as does the productivity. Table 0-1 summarizes the re-alignment of a verification environment based on SVA.

Though SVA interacts with the design signals directly, it can be used very effectively to share information with the testbenches. By sharing information dynamically during a simulation, very efficient reactive testbench environments can be developed. The completeness of the verification process can be measured more effectively by combining the code coverage and the functional coverage information collected during simulation.

Table 0-1. New verification environment

	Testbench	**SVA**
Before SVA	Stimulus generation Protocol checking Data checking Protocol coverage Test plan coverage	N/A
After SVA	Stimulus generation Data checking Test plan coverage	Protocol checking Protocol coverage

The book will introduce the SVA language, its use model and its benefits in an elaborate fashion with examples. It will show how to find bugs early by writing good quality assertions. Real design examples and the process of writing assertions to verify the design will be discussed. Measuring functional coverage on real designs and also how to use the functional coverage information dynamically to create more sophisticated testbenches will be discussed. Coding guidelines and simulation methodology practices will be discussed wherever relevant.

Chapter 1

INTRODUCTION TO SVA

Understanding the Syntax

1.1 What is an Assertion?

An assertion is a description of a property of the design.

- If a property that is being checked for in a simulation does not behave the way we expect it to, the assertion fails.
- If a property that is forbidden from happening in a design happens during simulation, the assertion fails.

A list of the properties can be inferred from the functional specification of a design and can be converted into assertions. These assertions can be continuously monitored during functional simulations. The same assertions can also be re-used for verifying the design using formal techniques. Assertions, also known as monitors or checkers, have been used as a form of debugging technique for a very long time in the design verification process. Traditionally, they are written in a procedural language like Verilog. They can also be written in PLI and C/C++ programs. The following code shows a simple mutually asserted condition check written in Verilog, wherein signal "a" and signal "b" cannot be high at the same time. If they are, an error message is displayed.

```
`ifdef ma
if(a & b)
$display
```

```
("Error:Mutually asserted check failed\n");
`endif
```

This kind of a monitor is included only as part of the simulation and hence is included in the design environment only on a need basis. This can be accomplished with the **`ifdef** construct which enables conditional compilation of Verilog code.

1.2 Why use SystemVerilog Assertions (SVA)?

While Verilog language can be used to write certain checks easily, it has a few disadvantages.

1. Verilog is a procedural language and hence, does not have good control over time.
2. Verilog is a verbose language. As the number of assertions increase, it becomes very difficult to maintain the code.
3. The procedural nature of the language makes it difficult to test for parallel events in the same time period. In some cases, it is also possible that a Verilog checker might not capture all the triggered events.
4. Verilog language has no built-in mechanism to provide functional coverage data. The user has to produce this code.

SVA is a declarative language and is perfectly suited for describing temporal conditions. The declarative nature of the language gives excellent control over time. The language itself is very concise and is very easy to maintain. SVA also provides several built-in functions to test for certain design conditions and also provides constructs to collect functional coverage data automatically.

Example 1.1 shows a checker written both in Verilog and SVA. The checker verifies that if signal "a" is high in the current clock cycle, then signal "b" should be high within 1 to 3 clock cycles. Figure 1-1 shows the waveform of a sample simulation of the signals "a" and "b."

Example 1.1 Sample assertion written in Verilog and SVA

```
// Sample Verilog checker

always @(posedge a)
begin
```

```
repeat (1) @(posedge clk);
  fork: a_to_b

     begin
     @(posedge b)
     $display
     ("SUCCESS: b arrived in time\n", $time);
     disable a_to_b;
     end

     begin
     repeat (3) @(posedge clk);
     $display
     ("ERROR:b did not arrive in time\n", $time);
     disable a_to_b;
     end

   join
end

// SVA Checker

a_to_b_chk:
assert property
@(posedge clk) $rose(a) |-> ##[1:3] $rose(b));
```

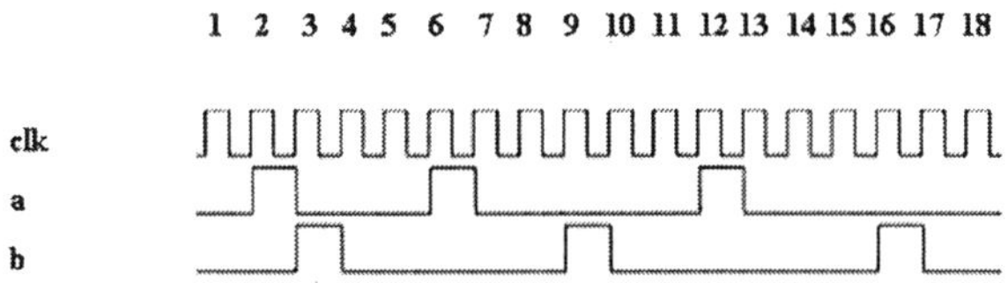

Figure 1-1. Waveform for sample assertion

Example 1.1 shows the advantages of SVA very clearly. SVA syntax is discussed in detail in this chapter. The checker represents a very simple protocol. It can be written in one line in SVA, although the same protocol description takes several lines in Verilog. Also, the error and success conditions need to be defined in Verilog explicitly, whereas the failure will automatically display an error message in SVA. Results of a sample simulation are shown below.

```
SUCCESS: b arrived in time 127
vtosva.a_to_b_chk:
started at 125s succeeded at 175s

SUCCESS: b arrived in time 427
vtosva.a_to_b_chk:
started at 325s succeeded at 475s

ERROR: b did not arrive in time 775
vtosva.a_to_b_chk:
started at 625s failed at 775s
        Offending '$rose(b)'
```

1.3 SystemVerilog Scheduling

The SystemVerilog language is defined to be an event based execution model. In each time slot, many events are scheduled to happen. This list of events follows the algorithm specified by the standard. By following this algorithm, the simulators can avoid any inconsistencies in the interactions between the design and testbench. There are three regions that are involved in the evaluation and execution of the assertions.

Preponed – Values are sampled for the assertion variables in this region. In this region, a net or variable cannot change its state. This allows the sampling of the most stable value at the beginning of the time slot.

Observed – All the property expressions are evaluated in this region.

Reactive – The pass/fail code from the evaluation of the properties are scheduled in this region.

Figure 1-2 shows a simplified SystemVerilog event schedule flow chart. To understand the SystemVerilog scheduling algorithm thoroughly, please refer to the SystemVerilog 3.1a LRM [1].

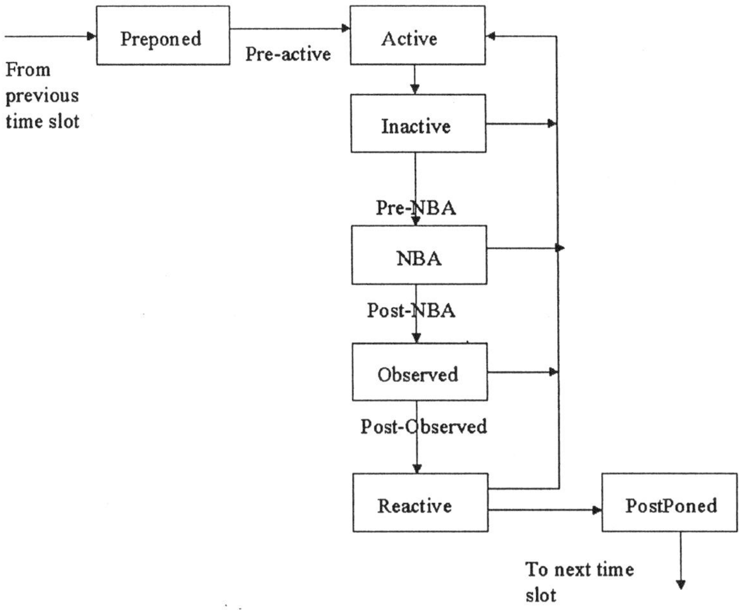

Figure 1-2. Simplified SV event schedule flow chart

1.4 SVA Terminology

There are two types of assertions defined in the SystemVerilog language: Concurrent assertions and Immediate assertions.

1.4.1 Concurrent assertions

- Based on clock cycles.
- Test expression is evaluated at clock edges based on the sampled values of the variables involved.
- Sampling of variables is done in the "preponed" region and the evaluation of the expression is done in the "observed" region of the scheduler.
- Can be placed in a procedural block, a module, an interface or a program definition.
- Can be used with both static (formal) and dynamic verification (simulation) tools.

A sample concurrent assertion is shown below.

```
a_cc: assert property(@(posedge clk)
                              not (a && b));
```

Figure 1-3 shows the results of the concurrent assertion a_cc. All successes are shown with an up arrow and all failures are shown with a down arrow. The key concept in this example is that the property is being verified on every positive edge of the clock irrespective of whether or not signal "a" and signal "b" changes.

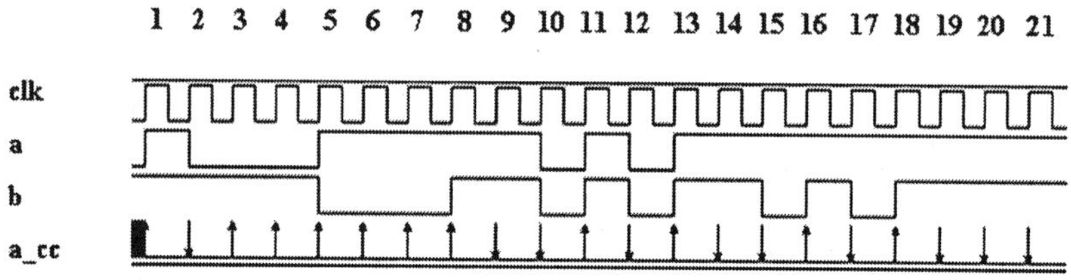

Figure 1-3. Waveform for a sample concurrent assertion

1.4.2 Immediate assertions

- Based on simulation event semantics.
- Test expression is evaluated just like any other Verilog expression within a procedural block. These are not temporal in nature and are evaluated immediately.
- Have to be placed in a procedural block definition.
- Used only with dynamic simulation.

A sample immediate assertion is shown below.

```
always_comb
begin
   a_ia: assert (a && b);
end
```

The immediate assertion a_ia is written as part of a procedural block and it follows the same event schedule of signal "a" and "b." The always block executes if either signal "a" or signal "b" changes. The keyword that differentiates the immediate assertion from the concurrent assertion is "**property.**" Figure 1-4 shows the results of the immediate assertion a_ia.

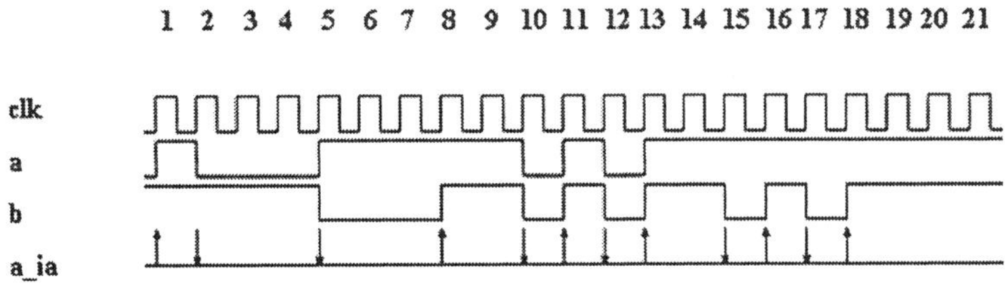

Figure 1-4. Waveform for a sample immediate assertion

1.5 Building blocks of SVA

In any design model, the functionality is represented by the combination of multiple logical events. These events could be simple boolean expressions that get evaluated on the same clock edge or could be events that evaluate over a period of time involving multiple clock cycles. SVA provides a key word to represent these events called "**sequence.**" The basic syntax of a **sequence** is as follows.

```
sequence name_of_sequence;
   < test expression>;
endsequence
```

A number of sequences can be combined logically or sequentially to create more complex sequences. SVA provides a key word to represent these complex sequential behaviors called "**property.**" The basic syntax of a **property** is as follows.

```
property name_of_property;
   < test expression >; or
   < complex sequence expressions >;
endproperty
```

The property is the one that is verified during a simulation. It has to be asserted to take effect during a simulation. SVA provides a key word called "**assert**" to check the property. The basic syntax of an **assert** is as follows.

```
assertion_name: assert property(property_name);
```

The steps involved in the creation of a SVA checker are shown in Figure 1-5.

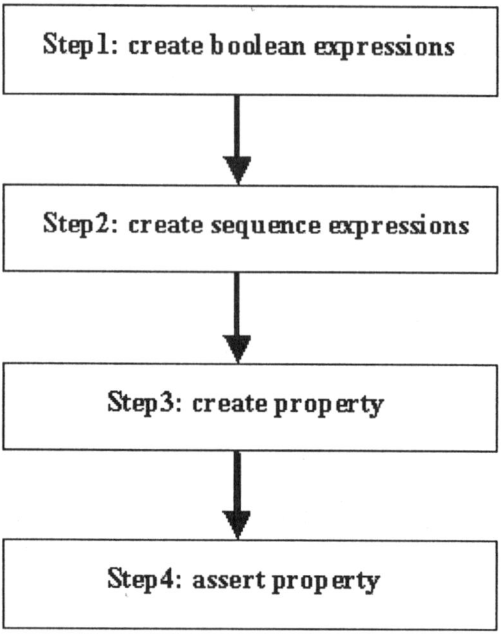

Figure 1-5. SVA Building blocks

1.6 A simple sequence

Sequence s1 checks that the signal "a" is high on every positive edge of the clock. If signal "a" is not high on any positive clock edge, the assertion will fail. Note that "a" is the same as "a==1'b1."

```
sequence s1;
  @(posedge clk) a;
endsequence
```

Figure 1-6 shows a sample waveform for signal "a" and how sequence s1 responds to this signal during simulation. Signal "a" goes to zero on the positive edge of clock cycle 7. This change in value is sampled in clock cycle 8. Since concurrent assertions use the values sampled in the "preponed" region of the scheduler, in clock cycle 7, the most stable value of signal "a" sampled by the sequence s1 is 1. Hence, the sequence succeeds. In clock cycle 8, the sampled value of signal "a" is a 0 and hence the sequence fails. A success is denoted with an arrow pointing up and a failure is denoted with an arrow pointing down. Table 1-1 summarizes the sampled values of signal "a" on each clock cycle up to clock cycle 15.

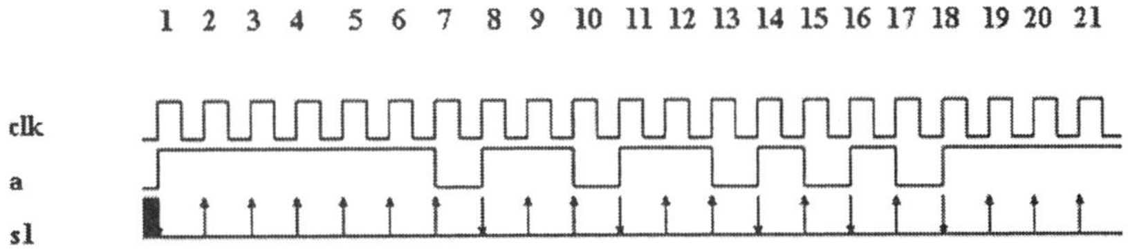

Figure 1-6. Waveform for simple sequence s1

Table 1-1. Evaluation table for sequence s1

Clock tick	Sampled value of signal "a"
1	0
2	1
3	1
4	1
5	1
6	1
7	1
8	0
9	1
10	1
11	0
12	1
13	1
14	0
15	1

1.7 Sequence with edge definitions

In sequence s1, the logical value of the signal was used. SVA also has built-in edge expressions that let the user monitor the transition of signal value from one clock cycle to the next. This allows one to check for the edge sensitivity of signals. Three of these useful built-in functions are shown below.

$rose (*boolean expression or signal_name*)

- This returns true if LSB of signal/expression changed to 1

$fell (*boolean expression or signal_name*)

- This returns true if LSB of signal/expression changed to 0

$stable (*boolean expression or signal_name*)

- This returns true if the value of the expression did not change

Sequence s2 checks that the signal "a" transitions to a value of 1 on every positive edge of the clock. If the transition does not occur, the assertion will fail.

```
sequence s2;
  @(posedge clk) $rose(a);
endsequence
```

Figure 1-7 shows how sequence s2 responds to the transition of signal "a." Marker 1 shows the first success of sequence s2. At clock cycle 1, the value of signal "a" goes from 0 to 1. At this clock, the sampled value of signal "a" within the sequence is 0. Before clock cycle 1, there is no history for signal "a" and hence the value is assumed to be "x." A transition of value from x to 0 is not a rising edge and hence the sequence fails. At clock cycle 2, the sampled value of signal "a" within the sequence is 1. A transition of value from 0 to 1 is a rising edge and hence, the sequence s2 succeeds in clock cycle 2. Another success is shown with marker 2 at clock cycle 9. Table 1-2 summarizes the transition of signal "a" over time until clock cycle 9 and how the sequence samples and updates the values.

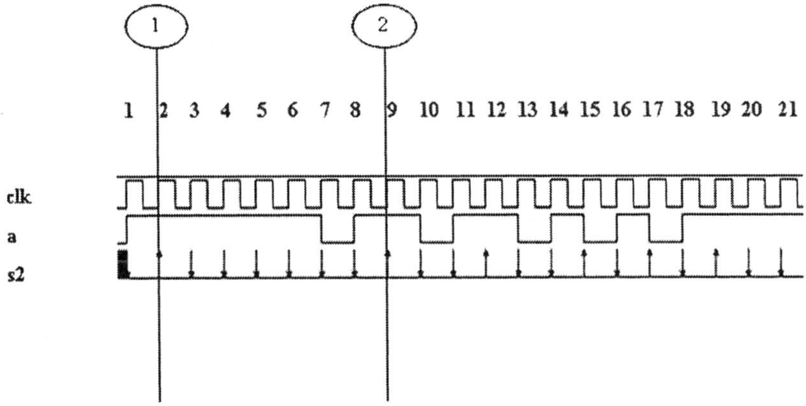

Figure 1-7. Waveform for simple sequence with edge definition

Table 1-2. Evaluation table for sequence s2

Clock Tick	Sampled value of "a" from the previous cycle	Sampled value of "a" in the current cycle	Sequence s2 - status
1	X	0	Fail
2	0	1	Success
3	1	1	Fail
4	1	1	Fail
5	1	1	Fail
6	1	1	Fail
7	1	1	Fail
8	1	0	Fail
9	0	1	Success

1.8 Sequence with logical relationship

Sequence s3 checks that on every positive edge of the clock, either signal "a" or signal "b" is high. If both the signals are low, the assertion will fail.

```
sequence s3;
  @(posedge clk)  a || b;
endsequence
```

Figure 1-8 shows how the sequence s3 responds to signal "a" and "b." Marker 1 shows that at clock cycle 12, the sampled values of both signals "a" and "b" are 0 and hence the sequence fails. The same is true for clock cycle 17 shown by marker 2. In all other clock cycles, either signal "a" or signal "b" has a value of 1 and hence the sequence succeeds in those clock cycles.

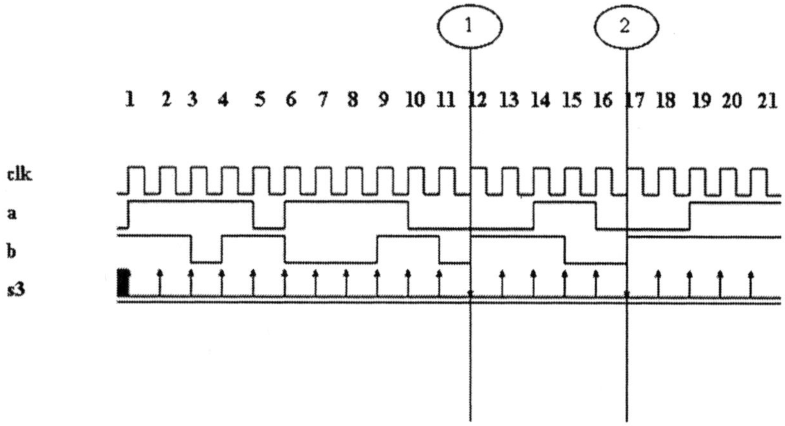

Figure 1-8. Waveform for sequence s3

1.9 Sequence Expressions

By defining formal arguments in a sequence definition, the same sequence can be re-used on other signals of a design that have similar behavior. For example, we can define a sequence as follows.

```
sequence s3_lib (a, b);
   a || b;
endsequence
```

The generic sequence s3_lib can be re-used on any two signals. For example, say we have two signals "req1" and "req2" and one of them should be asserted on the positive edge of a clock. We can write a sequence as follows.

```
sequence s3_lib_inst1;
   S3_lib(req1, req2);
```

```
endsequence
```

Some of the common properties that are normally present in designs can be developed as a library and re-used. For example, one-hot state machine checks, parity checks, etc. are good candidates for a checker library.

1.10 Sequences with timing relationship

Simple boolean expressions are checked on every clock edge. In other words, they are simple combinational checks. A lot of times, we are interested in checking events that take several clock cycles to complete. These are called "**sequential checks.**" ***In SVA, clock cycle delays are represented by a "##" sign. For example, ##3 means 3 clock cycles.***

Sequence s4 checks for the signal "a" being high on a given positive edge of the clock. If signal "a" is not high, then the sequence fails. If signal "a" is high on any given positive edge of clock, then signal "b" should be high 2 clock cycles after that. If signal "b" is not asserted after 2 clock cycles, the assertion fails. Note that the sequence begins when signal "a" is high on a positive edge of the clock.

```
sequence s4;
  @(posedge clk)  a ##2 b;
endsequence
```

Figure 1-9 shows how sequence s4 responds in a simulation. Table 1-3 summarizes the sampled values of signal "a" and signal "b" on every clock cycle.

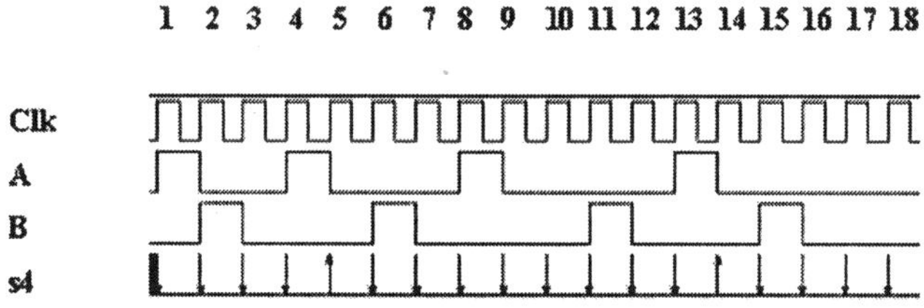

Figure 1-9. Waveform for sequence s4

Unlike the examples from the previous section, note that the start and end time of sequence s4 are not the same. If signal "a" is not high on any given clock cycle, then the sequence starts and fails on the same clock cycle. If signal "a" is high, then the sequence starts. The sequence succeeds after 2 clock cycles if signal "b" is high (clock 5 and clock 14). On the other hand, if signal "b" is not high after 2 clock cycles, then the sequence fails. Note that the success of a sequence is always represented in the figure at the starting point of the sequence.

Table 1-3. Evaluation table for sequence s4

Clock tick	Sampled value of "a"	Sampled value of "b"	Valid start of s4	S4 status
1	0	0	No	Fail
2	1	0	Yes	Fail (start at 2, end at 4)
3	0	1	No	Fail
4	0	0	No	Fail
5	1	0	Yes	Success (start at 5, end at 7)
6	0	0	No	Fail
7	0	1	No	Fail
8	0	0	No	Fail
9	1	0	Yes	Fail (start at 9, end at 11)
10	0	0	No	Fail
11	0	0	No	Fail
12	0	1	No	Fail
13	0	0	No	Fail
14	1	0	Yes	Success (start at 14, end at 16)
15	0	0	No	Fail
16	0	1	No	Fail
17	0	0	No	Fail

1.11 Clock definitions in SVA

A sequence or a property does not do anything by itself in a simulation. They have to be asserted to take effect as shown below.

```
sequence s5;
  @(posedge clk)  a ##2 b;
endsequence

property p5;
  s5;
endproperty

a5 : assert property(p5);
```

Note that the clock is specified in the sequence s5. While this is one way of relating a check to a clock there are also other ways of doing it. A clock can be specified in a sequence, in a property or even in an assert statement. The following code shows the clock defined in the property definition p5a.

```
sequence s5a;
  a ##2 b;
endsequence

property p5a;
  @(posedge clk)  s5a;
endproperty

a5a : assert property(p5a);
```

In general, it is a good idea to define the clocks in property definitions and keep the sequences independent of the clocks. This will help increase the re-use of the basic sequence definitions.

A separate property definition is not needed to assert a sequence. Since the assert statement calls a property, the expression to be checked can be called from the assert statement directly as shown below in assertion a5b.

```
sequence s5b;
  a ##2 b;
endsequence
```

```
a5b : assert property(@(posedge clk) s5b);
```

While we can call a sequence with a clock definition from within the assert statement, calling a property with a clock definition from within the assert statement is not allowed. This coding style is shown below in assertion a5c.

```
a5c : assert property(@(posedge clk) p5a); // Not allowed
```

1.12 Forbidding a property

In all the examples shown so far, the property is checking for a true condition. A property can also be forbidden from happening. In other words, we expect the property to be false always. If the property is true, the assertion fails.

Sequence s6 checks that if signal "a" is high on a given positive edge of the clock, then after 2 clock cycles, signal "b" shall not be high. The keyword "**not**" is used to specify that the property should never be true.

```
sequence s6;
 @(posedge clk)  a ##2 b;
endsequence

property p6;
 not s6;
endproperty

a6 : assert property(p6);
```

Figure 1-10 shows how the checker a6 responds in a simulation. Note that the checker fails on two occasions (clock 5 and clock 14) as shown by markers 1 and 2. In both these clock cycles, the sequence that was forbidden happened and hence asserted a failure.

On the other hand, the checker passes on two occasions when there is a valid signal "a" (clock 2 and clock 9). For the checks that began in these clock cycles, signal "b" does not go high after two clock cycles and hence the checker succeeded. All other clock cycles wherein signal "a" was not high succeeded automatically. Table 1-4 summarizes the sampled values of signal "a" and signal "b" on each clock cycle.

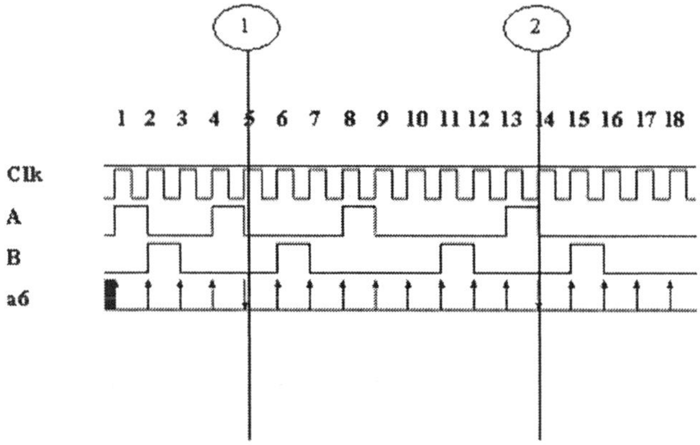

Figure 1-10. Waveform of SVA checker forbidding a property

Table 1-4. Evaluation table for property p6

Clock tick	Sampled value of "a"	Sampled value of "b"	Valid start of s6	a6 status
1	0	0	Yes	Success (same clock)
2	1	0	Yes	Success (start at 2, end at 4)
3	0	1	Yes	Success (same clock)
4	0	0	Yes	Success (same clock)
5	1	0	Yes	Fail (start at 5, end at 7)
6	0	0	Yes	Success (same clock)
7	0	1	Yes	Success (same clock)
8	0	0	Yes	Success (same clock)
9	1	0	Yes	Success (start at 9, end at 11)
10	0	0	Yes	Success (same clock)
11	0	0	Yes	Success (same clock)
12	0	1	Yes	Success (same clock)
13	0	0	Yes	Success (same clock)
14	1	0	Yes	Fail (start at 14, end at 16)
15	0	0	Yes	Success (same clock)
16	0	1	Yes	Success (same clock)
17	0	0	Yes	Success (same clock)

1.13 A simple action block

The SystemVerilog language is defined such that, every time an assertion check fails, the simulator is expected to print out an error message by default. The simulator need not print anything upon a success of an assertion. A user can also print a custom error or success message using the "**action block**" in the assert statement. The basic syntax of an **action block** is shown below.

```
assertion_name :
     assert property(property_name)
                <success message> ;
          else
                <fail message>;
```

The checker a7 shown below uses simple display statements in the action block to print successes and failures.

```
property p7;
  @(posedge clk)  a ##2 b;
endproperty

a7 : assert property(p7)
   $display("Property p7 succeeded\n");
   else
   $display("Property p7 failed\n");
```

The action block is not just limited to displaying success and failure. It can be used for other applications such as controlling the simulation environment and gathering functional coverage data. These topics will be discussed in detail in Chapter 2.

1.14 Implication operator

In the property p7, the following can be noticed.

1. The property looks for a valid start of the sequence on every positive edge of the clock. In this case, it looks for signal "a" to be high on every positive clock edge.
2. If signal "a" is not high on any given positive clock edge, an error is issued by the checker. This is not a valid error message since we are not interested in just checking for a specific level on signal "a." This

error just means that we did not get a valid starting point for the checker at this clock. While these errors are benign, they can log a lot of error messages over time, since the check is performed on every clock edge. To avoid these errors, some kind of gating technique needs to be defined, which will ignore the check if a valid starting point is not present.

SVA provides a technique to achieve this goal. This technique is called "**Implication.**" Implication is equivalent to an if-then structure. The left hand side of the implication is called the "**antecedent**" and the right hand side is called the "**consequent.**" The antecedent is the gating condition. If the antecedent succeeds, then the consequent is evaluated. If the antecedent does not succeed, then the property is assumed to succeed by default. This is called a "vacuous success." While implication avoids unnecessary error messages, it can produce vacuous successes. ***The implication construct can be used only with property definitions. It cannot be used in sequences.***

There are 2 types of implication: Overlapped implication and Non-overlapped implication.

1.14.1 Overlapped implication

Overlapped implication is denoted by the symbol |->. If there is a match on the antecedent, then the consequent expression is evaluated in the same clock cycle. A simple example is shown below in property p8. This property checks that, if signal "a" is high on a given positive clock edge, then signal "b" should also be high on the same clock edge.

```
property p8;
  @(posedge clk)  a |-> b;
endproperty

a8 : assert property(p8);
```

Figure 1-11 shows how the assertion a8 responds in a simulation. Table 1-5 summarizes the sampled values of signal "a" and signal "b" and the status of the assertion. There are 3 types of results shown in the table. A real success is one where a valid high on signal "a" was detected, and at the same clock edge a valid high on signal "b" was detected. A vacuous success is one where signal "a" was not high and the assertion succeeded by default. A failure is one where a valid high on signal "a" was detected and at the same clock edge a valid high on signal "b" was not detected high.

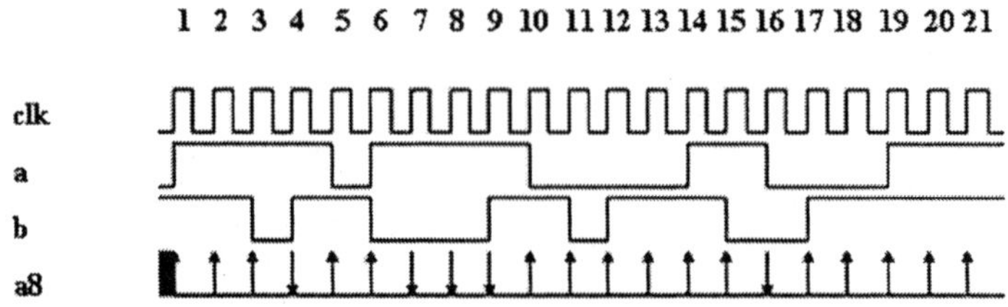

Figure 1-11. Waveform for property p8

Table 1-5. Evaluation table for property p8

Clock Tick	Sampled value of "a"	Sampled value of "b	A8 status
1	0	1	Vacuous success
2	1	1	Real Success
3	1	1	Real Success
4	1	0	Fail
5	1	1	Real Success
6	0	1	Vacuous success
7	1	0	Fail
8	1	0	Fail
9	1	0	Fail

1.14.2 Non-overlapped implication

Non-overlapped implication is denoted by the symbol |=>. If there is a match on the antecedent, then the consequent expression is evaluated in the next clock cycle. A delay of one clock cycle is assumed for the evaluation of the consequent expression. A simple example is shown below in property p9. This property checks that, if signal "a" is high on a given positive clock edge, then signal "b" should be high on the next clock edge.

```
property p9;
 @(posedge clk)  a |=> b;
endproperty

a9 : assert property(p9);
```

Figure 1-12 shows how the assertion a9 responds in a simulation. Table 1-6 summarizes the sampled values of signal "a" and signal "b" and the status of the assertion. Note that this assertion starts in the current clock cycle and succeeds in the next clock cycle if it is a real success. Similarly, if there is a valid start for the property (high on signal "a"), the property fails in the next clock cycle if signal "b" is not high in that clock cycle.

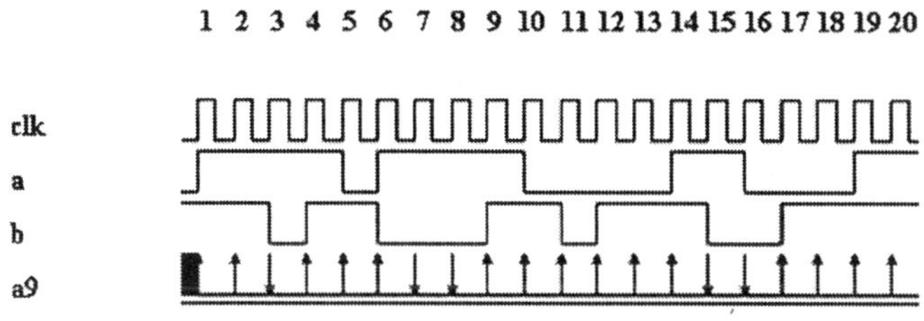

Figure 1-12. Waveform for property p9

Table 1-6. Evaluation table for property p9

Clock Tick	Sampled value of "a"	Sampled value of "b	a9 status
1	0	1	Vacuous success
2	1	1	Real success (start at 2, end at 3)
3	1	1	Fail (start at 3, end at 4)
4	1	0	Real success (start at 4, end at 5)
5	1	1	Real success (start at 5, end at 6)
6	0	1	Vacuous success
7	1	0	Fail (start at 7, end at 8)
8	1	0	Fail (start at 8, end at 9)
9	1	0	Real success (start at 9, end at 10)

1.14.3 Implication with a fixed delay on the consequent

Property p10 checks that if signal "a" is high in a given positive clock edge, then signal "b" should be high after 2 clock cycles. A similar check was shown before without the use of the implication operator. By using the implication, all the false errors are removed. A check for the consequent (signal "b") is performed only if there is a valid start for the property (high on signal "a"). Figure 1-13 shows a sample simulation of the property p10. Table 1-7 summarizes the sampled values of the signals involved in property p10.

```
property p10;
  @(posedge clk)  a |-> ##2 b;
endproperty

a10 : assert property(p10);
```

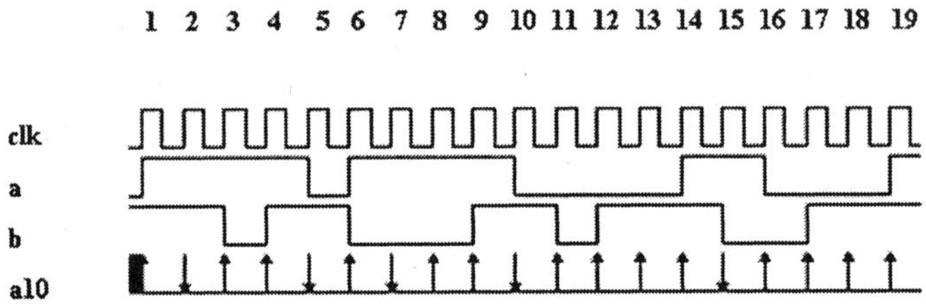

Figure 1-13. Waveform for property p10

Table 1-7. Evaluation table for property p10

Clock Tick	Sampled value of "a"	Sampled value of "b	a10 status
1	0	1	Vacuous success
2	1	1	Fail (start at 2, end at 4)
3	1	1	Success (start at 3, end at 5)
4	1	0	Success (start at 4, end at 6)
5	1	1	Fail (start at 5, end at 7)
6	0	1	Vacuous success
7	1	0	Fail (start at 7, end at 9)
8	1	0	Success (start at 8, end at 10)
9	1	0	Success (start at 9, end at 11)

1.14.4 Implication with a sequence as an antecedent

Property p10 has a signal in the antecedent position. It is also possible to have a sequence definition in the antecedent. In this case, a check for the consequent sequence or Boolean expression is performed only if the sequence in the antecedent succeeds. Sequence s11a checks that in any given positive clock edge, if signal "a" and signal "b" are detected to be high, then one clock cycle later, signal "c" should be high. Sequence s11b checks that, after 2 clock cycles from the current positive edge of the clock, signal "d" should be low. The final property checks that, if sequence s11a succeeds, then a check for sequence s11b is performed. If a valid sequence s11a is not detected, then the sequence s11b is not checked for and the property succeeds vacuously.

```
sequence s11a;
  @(posedge clk)   (a && b) ##1 c;
endsequence

sequence s11b;
  @(posedge clk)   ##2  !d;
endsequence

property p11;
  s11a |-> s11b;
endpeoperty

a11 : assert property(p11);
```

Figure 1-14 shows how the assertion a11 behaves in a simulation. The markers 1s and 1e show the start and end of a successful property evaluation. The markers 2s and 2e show the start and end of a failure. At clock cycle 11, both signal "a" and signal "b" are detected high. In clock cycle 12, signal "c" is high and hence the antecedent of the implication succeeds. This means that, 2 clock cycles from now, which is clock cycle 14, signal "d" should be low. But in the sample waveform signal "d" is a high and hence the property fails.

All the vacuous successes are shown with a simple straight line. The markers 3s and 3e show the start and end of a successful property evaluation. The expression "a **&&** b" is evaluated to be true in clock cycle 17 and one clock cycle later, the signal "c" is high, as expected. Hence, at clock cycle 18, the sequence s11a succeeds. The signal "d" is expected to be low 2 clock cycles from here and it is low as expected. Hence, the property succeeds at clock cycle 20.

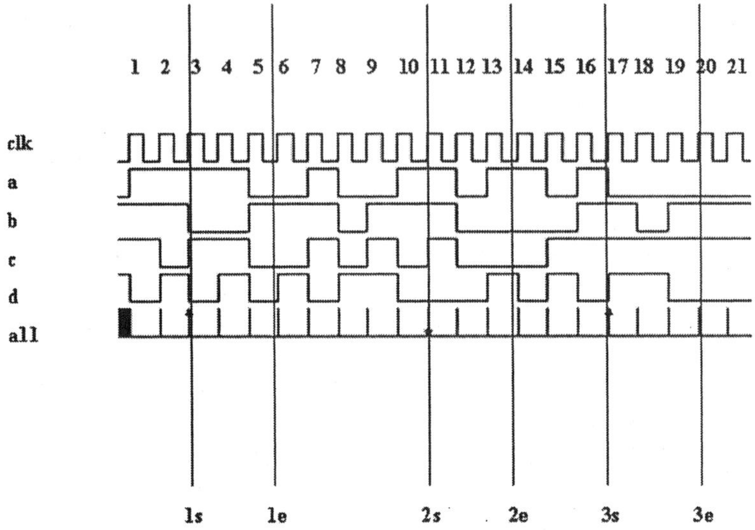

Figure 1-14. Waveform for property p11

1.15 Timing windows in SVA Checkers

So far, the examples shown with delays have a fixed delay greater than 0. In the next few examples, different ways of specifying delays will be discussed.

Property p12 checks whether the boolean expression "a **&&** b" is true on any given positive edge of the clock. If it is true, then within 1 to 3 clock cycles, the signal "c" should be high. SVA allows specifying a timing window for the consequent to match. The value specified in the left hand side of the timing window should be less than the value specified in the right hand side of the timing window. The left hand side can also have a value of 0. If 0 is specified in the left hand side, it means that the consequent must be checked starting from the same clock edge at which the antecedent succeeded.

```
property p12;
  @(posedge clk)  (a && b) |->  ##[1:3] c;
endproperty

a12 : assert property(p12);
```

Figure 1-15 shows how the property p12 responds in a simulation. Whenever a timing window is specified, multiple threads get kicked off for all possible matches in every clock edge. The property gets executed as three separate threads as follows.

```
(a && b) |-> ##[1] c or
(a && b) |-> ##[2] c or
(a && b) |-> ##[3] c
```

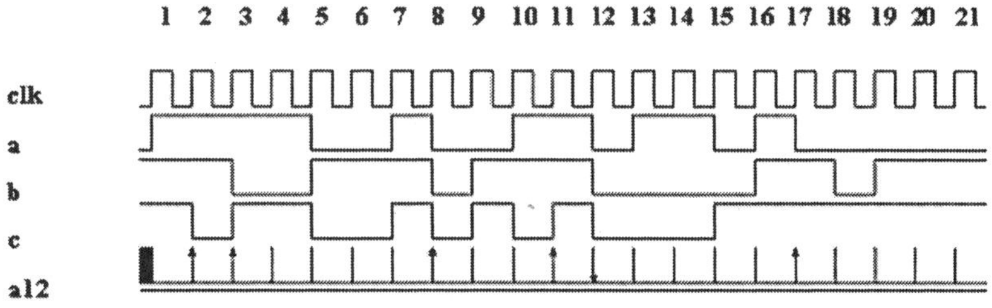

Figure 1-15. Waveform for p12

The property has 3 chances to succeed. All the three threads have the same starting point but the first thread that succeeds will make the property succeed. ***Also note that, there can be only one valid start on any give positive edge of the clock, but there can be multiple valid endings.*** This happens due to the fact that each valid start has 3 possible chances to succeed.

Table 1-8 summarizes the sampled values of all signals involved in the evaluation of the property. On a given positive clock edge, if signal "a" and signal "b" are both not high, then the property succeeds vacuously. On the other hand, on a given positive clock edge, if signal "a" and signal "b" are both high, then there is a valid start for the property. If signal "c" is not detected high in the next 1 to 3 clock cycles, the property fails.

Note that there is a valid start of the property detected on both clock cycle 2 and 3. Both of these valid starts succeed in clock cycle 4. The check that started at clock cycle 2 detected a high on signal "c" after 2 clock cycles. The check that started at clock cycle 3 detected a high on signal "c" after 1 clock cycle. Both of these are valid conditions and hence they succeed. There is also a valid start on clock cycle 12. The property checks for a high on signal "c" on clock cycles 13, 14 and 15. Since signal "c" remained low in all three possible clock cycles, the check failed.

Table 1-8. Evaluation table for property p12

Clock tick	Sampled value of "a"	Sampled value of "b"	Sampled value of "c"	Valid start of p12	a12 status
1	0	1	1	No	Vacuous success
2	1	1	1	Yes	Real Success (start at 2, end at 4)
3	1	1	0	Yes	Real Success (start at 3, end at 4)
4	1	0	1	No	Vacuous success
5	1	0	1	No	Vacuous success
6	0	1	0	No	Vacuous success
7	0	1	0	No	Vacuous success
8	1	1	1	Yes	Real Success (start at 8, end at 10)
9	0	0	0	No	Vacuous success
10	0	1	1	No	Vacuous success
11	1	1	0	Yes	Real Success (start at 11, end at 12)
12	1	1	1	Yes	Fail (start at 12, end at 15)
13	0	0	0	No	Vacuous success
14	1	0	0	No	Vacuous success
15	1	0	0	No	Vacuous success
16	0	0	1	No	Vacuous success
17	1	1	1	Yes	Real Success (start at 17, end at 18)

1.15.1 Overlapping timing window

Property p13 is similar to property p12. The main difference between the two is that the consequent of property p13 will be checked in the same clock edge in which the antecedent has a valid match.

```
Property p13;
  @(posedge clk)   (a && b)  |->   ##[0:2] c;
endproperty

a13 : assert property(p13);
```

Figure 1-16 shows how property p13 responds in a simulation. The main difference in the response when compared to property p12 is that, the valid start that happens in clock cycle 12 succeeds. This succeeds because of the overlap in checking. The value of signal "c" is detected high in the same clock edge as the valid match on the antecedent.

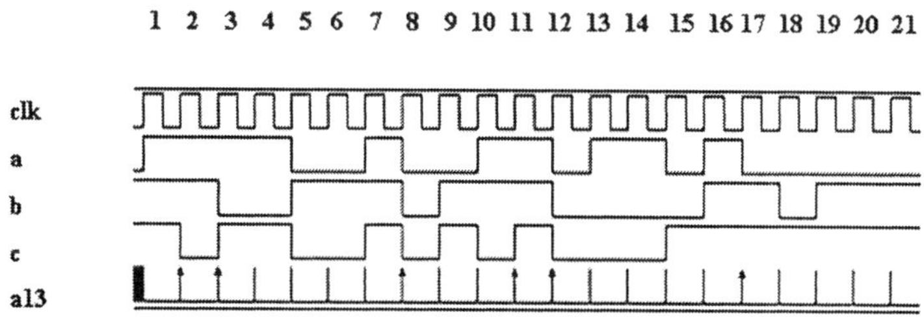

Figure 1-16. Waveform for property p13

1.15.2 Indefinite timing window

The upper limit of the timing window specified in the right hand side can be defined with a "$" sign which implies that there is no upper bound for timing. This is called the "**eventuality**" operator. The checker will keep checking for a match until the end of simulation. This is not a very efficient way of writing SVA since this has a huge impact on the simulation performance. It is best to always have a defined upper value in the timing window.

Property p14 checks that on a given positive edge of clock, signal "a" is high. If so, then signal "b" will be high eventually starting from the next clock cycle and after that, signal "c" will be high eventually starting at the same clock cycle in which signal "b" was high.

```
property p14;
  @(posedge clk)  a |->  ##[1:$] b ##[0:$] c;
endproperty

a14 : assert property(p14);
```

Figure 1-17 shows how property p14 reacts in a simulation. Table 1-9 summarizes the sampled values of the signals and the status of the assertion a14. Note that the real successes can take any number of clock cycles to finish. If there is a valid start and if either signal "b" or signal "c" does not match before the end of the simulation, these checks are reported as "**incomplete checks.**" Since overlap is allowed in the matching of signal "b" and signal "c," the whole check can finish in one clock cycle. Clock cycle 17 shows such a condition, wherein signal "a" was detected high on clock cycle 17 and both signal "b" and signal "c" were detected high on clock cycle 18.

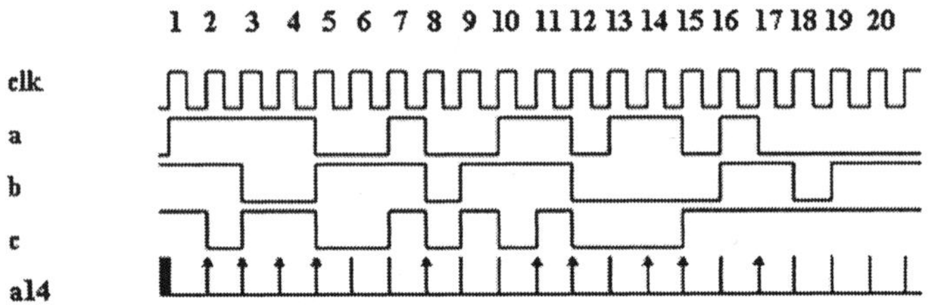

Figure 1-17. Waveform for property p14

Table 1-9. Evaluation table for p14

Clock tick	Sampled value of "a"	Sampled value of "b"	Sampled value of "c"	Valid start of p14	a14 status
1	0	1	1	No	Vacuous success
2	1	1	1	Yes	Real success (start at 2, end at 4)
3	1	1	0	Yes	Real success (start at 3, end at 8)

Clock tick	Sampled value of "a"	Sampled value of "b"	Sampled value of "c"	Valid start of p14	a14 status
4	1	0	1	Yes	Real success (start at 4, end at 8)
5	1	0	1	Yes	Real success (start at 5, end at 8)
6	0	1	0	No	Vacuous success
7	0	1	0	No	Vacuous success
8	1	1	1	Yes	Real success (start at 8, end at 10)
9	0	0	0	No	Vacuous success
10	0	1	1	No	Vacuous success
11	1	1	0	Yes	Real success (start at 11, end at 12)
12	1	1	1	Yes	Real success (start at 12, end at 17)
13	0	0	0	No	Vacuous success
14	1	0	0	Yes	Real success (start at 14, end at 17)
15	1	0	0	Yes	Real success (start at 15, end at 17)
16	0	0	1	No	Vacuous success
17	1	1	1	Yes	Real success (start at 17, end at 18)

1.16 The "ended" construct

The sequences defined so far use simple concatenation mechanism. In other words, multiple sequences were combined together over time by using the starting point of the sequence as the synchronization point. SVA provides another mechanism to concatenate sequences wherein the ending point of the sequence is used as a synchronization point. This is expressed by attaching the keyword "**ended**" to a sequence name. For example s.**ended** means the ending point of the sequence. The keyword **ended** stores a boolean value true or false depending on whether the sequence matched on that particular clock edge. ***This boolean value of the s.ended is available only in the same clock cycle.***

Sequence s15a and s15b are two 2 simple sequences that take more than 1 clock cycle to match. Property p15a checks that sequence s15a and sequence s15b match with a delay of one clock cycle in between them. Property p15b checks the same protocol but by using the keyword **ended**. In this case, the end point of the sequences does the synchronization. Since the endpoints are used, a delay of 2 clock cycles is defined between the 2 sequences.

```
sequence s15a;
 @(posedge clk) a ##1 b;
endsequence

sequence s15b;
 @(posedge clk) c ##1 d;
endsequence

property p15a;
 s15a |=> s15b;
endproperty

property p15b;
 s15a.ended |-> ##2 s15b.ended;
endproperty

a15a: assert property(p15a);
a15b: assert property(p15b);
```

Figure 1-18 shows how properties p15a and p15b react in a simulation. Table 1-10 summarizes the status of the assertions a15a and a15b. The first real success for assertion a15a happens at clock cycle 2. The check becomes active at clock cycle 2 when signal "a" is detected high. The check completes at clock cycle 5 when signal "d" is detected high. The first real success for assertion a15b occurs at clock cycle 3. The check becomes active at clock cycle 3 when the sequence s15a matches or in other words, signal "b" is detected high. The check completes at clock cycle 5 when the sequence s15b matches or in other words, when signal "d" is detected high.

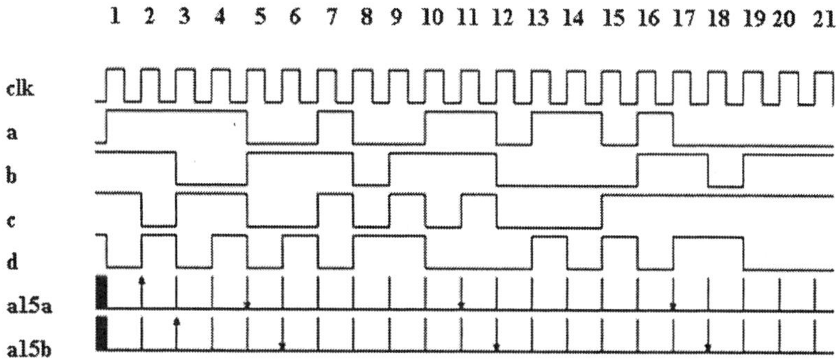

Figure 1-18. Waveform for SVA checker using "ended"

The first failure for assertion a15a happens at clock cycle 5. A valid starting point is detected when signal "a" is detected high on a given positive clock edge and is followed by a high on signal "b" one clock cycle later (clock cycle 6). This leads to checking the consequent and since signal "c" is not high after one clock cycle, the check fails at clock cycle 7.

The first failure for assertion a15b occurs at clock cycle 6. A valid starting point is detected when sequence s15a ends successfully at clock cycle 6. This leads to checking the consequent wherein a valid end point for sequence s15b is expected at clock cycle 8. Since signal "c" does not go high as expected at clock cycle 7, the end point value of the sequence is false and hence the check fails at clock cycle 8.

There are 2 different ways of writing the same check. The first method synchronizes the sequences based on the starting points of the sequences. The second method synchronizes the sequences based on the end points of the sequences.

Table 1-10. Evaluation table for SVA checker using "ended"

Clock tick	Sampled value of "a"	Sampled value of "b"	Sampled value of "c"	Sampled value of "d"	A15a status	A15b status
1	0	1	1	1	Vacuous success	Vacuous success
2	1	1	1	0	Real Success (start at 2, end at 5)	Vacuous success
3	1	1	0	1	Vacuous success	Real Success (start at 3, end at 5)
4	1	0	1	0	Vacuous success	Vacuous success
5	1	0	1	1	Fail (start at 5, end at 7)	Vacuous success
6	0	1	0	0	Vacuous success	Fail (start at 6, end at 8)
7	0	1	0	1	Vacuous success	Vacuous success
8	1	1	1	0	Vacuous success	Vacuous success
9	0	0	0	1	Vacuous success	Vacuous success
10	0	1	1	1	Vacuous success	Vacuous success
11	1	1	0	0	Fail (start at 11, end at 13)	Vacuous success
12	1	1	1	0	Vacuous success	Fail (start at 12, end at 14)
13	0	0	0	0	Vacuous success	Vacuous success
14	1	0	0	1	Vacuous success	Vacuous success
15	1	0	0	0	Vacuous success	Vacuous success
16	0	0	1	1	Vacuous success	Vacuous success
17	1	1	1	0	Fail (start at 17, end at 20)	Vacuous success

1.17 SVA Checker using parameters

SVA allows using parameters in the checkers just like Verilog. This gives great flexibility in creating re-usable properties. For example, the delay information between 2 signals can be parameterized within the checker and then the checker can be re-used in a similar situation elsewhere in the design with different timing relationships. Example 1.2 shows a checker defined with a default value for the parameter delay. If this checker is called within the design, it uses a delay of one clock cycle by default. The same checker can be re-used by over-writing the delay parameter value while instantiating the checker. In Example 1.2, module "top" has 2 instances of the "generic_chk" checker. Instance i1 overwrites the delay parameter as 2 clock cycles and instance i2 uses the default value of 1 clock cycle.

Example 1.2 Sample SVA checker using parameters

```
module generic_chk (input logic a, b, clk);

parameter delay = 1;

property p16;
  @(posedge clk) a |-> ##delay b;
endproperty

a16: assert property(p16);

endmodule

// call checker from the top level module

module top(.....);
logic clk, a, b, c, d;
.
.
generic_chk #(.delay(2)) i1 (a, b, clk);
generic_chk i2 (c, d, clk);
.
.
endmodule
```

Figure 1-19 shows how the 2 instances of checkers, i1 and i2, react to transitions in signals during a simulation.

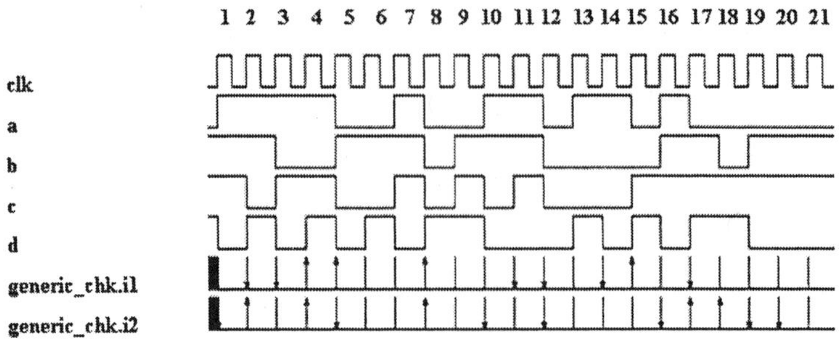

Figure 1-19. Waveform for SVA checker with parameters

1.18 SVA Checker using a select operator

SVA allows using logical operators within sequences and properties. Property p17 checks that, if signal "c" is high then the value of signal "d" is equal to the value of signal "a." If signal "c" is not detected high, then the value of signal "d" is equal to the value of signal "b." This is a combinational check and is performed on every positive edge of clock.

```
property p17;
  @(posedge clk) c ? d == a : d == b;
endproperty

a17: assert property(p17);
```

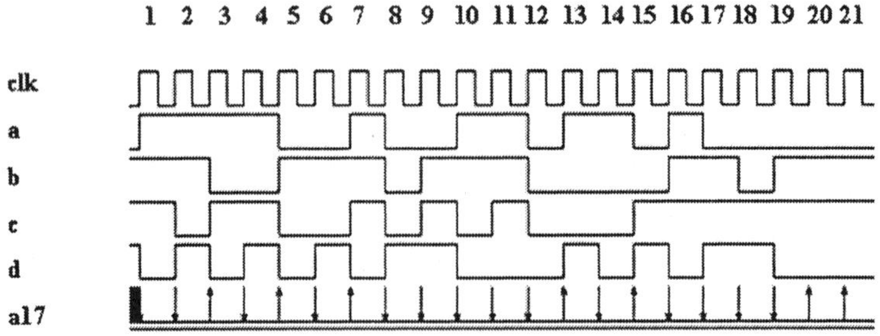

Figure 1-20. Waveform for SVA checker using select operator

Figure 1-20 shows how property p17 reacts in a simulation. Table 1-11 summarizes the sampled values of the respective signals and the status of the assertion a17. At clock cycle 1, signal "c" is detected high and hence, the check expects that signal "d" and signal "a" have the same value. But signal "d" is detected as high and signal "a" as low and hence the check fails.

Table 1-11. Evaluation table for SVA checker using select operator

Clock tick	Sampled value of "a"	Sampled value of "b"	Sampled value of "c"	Sampled value of "d"	a17 status
1	0	1	1	1	Fail
2	1	1	1	0	Fail
3	1	1	0	1	Success
4	1	0	1	0	Fail
5	1	0	1	1	Success
6	0	1	0	0	Fail
7	0	1	0	1	Success
8	1	1	1	0	Fail
9	0	0	0	1	Fail
10	0	1	1	1	Fail
11	1	1	0	0	Fail
12	1	1	1	0	Fail
13	0	0	0	0	Success
14	1	0	0	1	Fail
15	1	0	0	0	Success
16	0	0	1	1	Fail
17	1	1	1	0	Fail

1.19 SVA Checker using true expression

SVA checkers can be extended in time by using a **`true** expression. This represents a "don't care" condition and it extends the sequence by a clock cycle. This can be used when writing complex protocols wherein multiple properties are monitored and matched simultaneously.

Sequence s18a checks for a simple condition. Sequence s18a_ext checks for the same condition, but moves the match on this sequence by one clock cycle. This has an impact on when this sequence is used in the antecedent of a property. The end points of the 2 sequences are different and hence the clock cycle at which the consequent will be checked will vary.

Property p18 checks for a match on s18a.ended in the antecedent and 2 clock cycles later, checks for a match on s18b.ended. Property p18_ext checks for a match on s18a_ext.ended in the antecedent. The match on this is the same as the match on s18a.ended, but moved 1 clock cycle ahead. Hence, the consequent of property p18_ext needs to match after one clock cycle and not 2 clock cycles as defined in property p18. Both properties p18 and p18_ext check for the same condition, but they both have different matching points for their antecedents.

```
`define true 1

sequence s18a;
 @(posedge clk) a ##1 b;
endsequence

sequence s18a_ext;
 @(posedge clk) a ##1 b ##1 `true;
endsequence

sequence s18b;
 @(posedge clk) c ##1 d;
endsequence

property p18
 @(posdge clk) s18a.ended |-> ##2 s18b.ended;
endproperty

property p18_ext
 @(posdge clk) s18a_ext.ended |=> s18b.ended;
endproperty

a18: assert property(p18);
a18_ext: assert property(p18_ext);
```

Figure 1-21 shows how property p18 and p18_ext react in a simulation. It is clearly seen that the starting point of assertion a18_ext is delayed by one cycle when compared to that of the assertion a18.

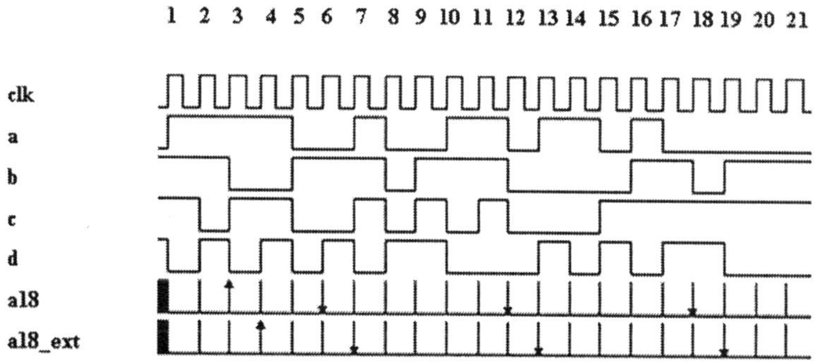

Figure 1-21. Waveform for SVA checker using `true expression

1.20 The "$past" construct

SVA provides a built in system task called **$past** that is capable of getting values of signals from previous clock cycles. By default, it provides the value of the signal from the previous clock cycle. The simple syntax of this construct is as follows.

$past (signal_name, number of clock cycles)

This task can be used effectively to verify that, the path taken by the design to get to the state in this current clock cycle is valid. Property p19 checks that in the given positive clock edge, if the expression (c **&&** d) is true, then 2 cycles before that, the expression (a **&&** b) was true.

```
Property p19;
  @(posedge clk) (c && d) |->
        ($past((a&&b), 2) == 1'b1);
endproperty

a19: assert property(p19);
```

Figure 1-22 shows how the property p19 reacts in a simulation. Table 1-12 summarizes the sampled values of the relevant signals and the status of the assertion a19. The assertion fails at clock cycle 1. At clock cycle 1, there is a valid start since both signal "c" and signal "d" are high. The consequent of the checker needs to compare the value of the expression (a **&&** b) 2 cycles before. This is not possible since there is no history for these signals before clock cycle 1 and hence the values are assumed to be "x." Hence, the checker fails at clock cycle 1.

The check has a real success at clock cycle 5. At clock cycle 5, there is a valid start since both signal "c" and signal "d" are high. The consequent checks that at clock cycle 3, the expression (a **&&** b) is true. As expected, at clock cycle 3, the signals "a" and "b" are detected high and hence the check succeeds.

The check fails at clock cycle 16. At clock cycle 16, there is a valid start, since both signal "c" and signal "d" are high. The consequent checks that at clock cycle 14, the expression (a **&&** b) is true. The signal "a" is detected high as expected and signal "b" is detected low. This makes the expression (a **&&** b) false and hence the check fails.

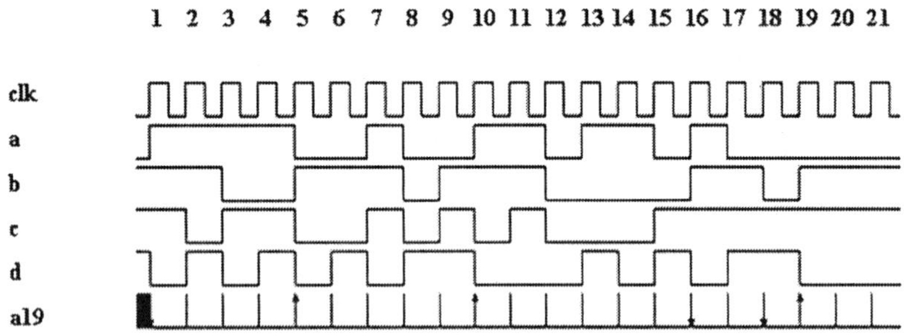

Figure 1-22. Waveform for SVA checker using "**$past**" construct

Table 1-12. Evaluation table for SVA checker using **$past** construct

Clock tick	**Sampled value of "a"**	**Sampled value of "b"**	**Sampled value of "c"**	**Sampled value of "d"**	**a19 status**
1	0	1	1	1	Fail
2	1	1	1	0	Vacuous success
3	1	1	0	1	Vacuous success
4	1	0	1	0	Vacuous success
5	1	0	1	1	Real Success
6	0	1	0	0	Vacuous success
7	0	1	0	1	Vacuous success

Clock tick	Sampled value of "a"	Sampled value of "b"	Sampled value of "c"	Sampled value of "d"	a19 status
8	1	1	1	0	Vacuous success
9	0	0	0	1	Vacuous success
10	0	1	1	1	Real Success
11	1	1	0	0	Vacuous success
12	1	1	1	0	Vacuous success
13	0	0	0	0	Vacuous success
14	1	0	0	1	Vacuous success
15	1	0	0	0	Vacuous success
16	0	0	1	1	Fail
17	1	1	1	0	Vacuous success

1.20.1 The $past construct with clock gating

The **$past** construct can be used with a gating signal. For example, on a given clock edge, the gating signal has to be true even before checking for the consequent condition. The simple syntax of a **$past** construct with a gating signal is as follows.

$past (signal_name, number of clock cycles, gating signal)

Property p20 is similar to the property p19. But the check is effective only if the gating signal "e' is valid on any given positive edge of the clock.

```
Property p20;
  @(posedge clk) (c && d) |->
               ($past((a&&b), 2, e) == 1'b1);
endproperty

a20: assert property(p20);
```

1.21 Repetition operators

If signal "start" is high on a given positive edge of the clock, then, starting from the next clock cycle, signal "a" stays high for 3 continuous clock cycles; one clock cycle after that, signal "stop" is high.

A sequence like this can be checked by the following SVA code.

```
@(posedge clk) $rose(start) |->
              ##1 a ##1 a ##1 a ##1 stop
```

Writing such a checker can get very verbose if signal "a" has to stay high for many cycles. Also, in this case, it is assumed that signal "a" stays high continuously. This protocol can get complex when we want to check if signal "a" stays high, not necessarily on three continuous clock cycles. In other words, signal "a" should repeat itself 3 times continuously or intermittently.

SVA language provides three different types of repetition operators: Consecutive repetition, go to repetition and non-consecutive repetition.

Consecutive repetition – This allows the user to specify that a signal or a sequence will match continuously for the number of clocks specified. A hidden delay of one clock cycle is assumed between each match of the signal. The simple syntax of consecutive repetition operator is shown below.

```
signal or sequence [*n]
```

"n" is the number of times the expression should match repeatedly.

For example a [*3] will expand to the following.

```
a ##1 a ##1 a
```

A sequence such as (a ##1 b) [*3] will expand as follows.

```
(a ##1 b) ##1 (a ##1 b) ##1 (a ##1 b)
```

Go to repetition – This allows the user to specify that an expression will match the number of times specified not necessarily on continuous clock cycles. The matches can be intermittent. The main requirement of a "go to" repeat is that the last match on the expression checked for repetition should happen in the clock cycle before the end of the entire sequence matching. The simple syntax of "go to" repetition operator is shown below.

```
Signal [->n]
```

Consider the following sequence.

```
Start ##1 a[->3] ##1 stop
```

It is required that there is a match on signal "a" (the third and final repetition of signal "a") just before the success of "stop." In other words, signal "stop" succeeds on the last clock cycle of the sequence match, and in the previous clock cycle, there should be a match on signal "a."

Non-consecutive repetition – This is very similar to "go to" repetition except that it does not require that the last match on the signal repetition happen in the clock cycle before the end the entire sequence matching. The simple syntax of a non-consecutive repetition operator is shown below.

```
Signal [=n]
```

Only expressions are allowed to repeat in "go to" and "non-consecutive" repetitions. Sequences are not allowed.

1.21.1 Consecutive repetition operator [*]

Property p21 checks that two clock cycles after a valid start, signal "a" stays high for 3 continuous clock cycles and two clock cycles after that, signal "stop" is high. One clock cycle later signal "stop" is low.

```
Property p21;
 @(posedge clk) $rose(start) |->
                   ##2 (a[*3]) ##2 stop ##1 !stop;
endproperty

a21: assert property(p21);
```

Figure 1-23 shows how property p21 reacts in a simulation. The waveform shows 2 failures and 1 real success. All other successes are vacuous.

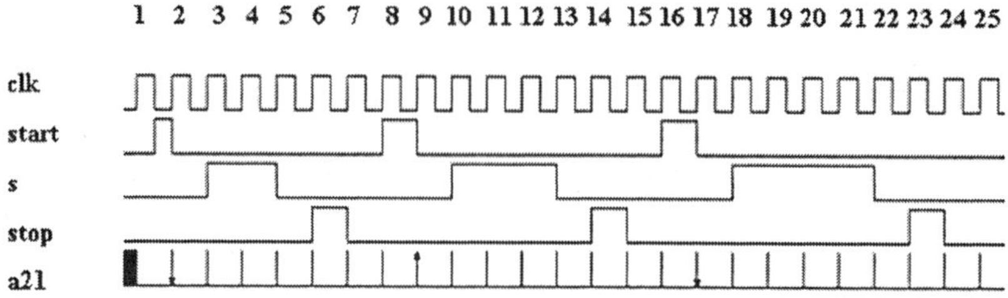

Figure 1-23. Waveform for SVA checker using consecutive repeat

Failure at clock cycle 2 – A valid start signal is detected at clock cycle 2. The checker then looks for signal "a" to be high on 3 continuous clock cycles starting from the positive clock edge of clock cycle 4. Signal "a" is detected high on clock cycle 4 and 5, but is detected low on clock cycle 6. Hence, the check fails. Note that the check started at clock cycle 2 and failed at clock cycle 6.

Success at clock cycle 9 - A valid start signal is detected at clock cycle 9. The checker then looks for signal "a" to be high for 3 continuous clock cycles starting from the positive clock edge of clock cycle 11. Signal "a" is detected high on clock cycle 11, 12 and 13 as expected. Two clock cycles later (at clock cycle 15) signal "stop" is high as expected. One clock cycle later the signal "stop" is detected low. Hence, the check succeeds. Note that the check started at clock cycle 9 and finished at clock cycle 16.

Failure at clock cycle 17 - A valid start signal is detected at clock cycle 17. The checker then looks for signal "a" to be high for 3 continuous clock cycles starting from the positive clock edge of clock cycle 19. Signal "a" is detected high on clock cycles 19, 20 and 21. The check now looks for a high on signal "stop" at clock cycle 23 but it is not there. Hence, the check fails. Note that signal "a" remained high for 4 clock cycles. The checker needs only 3 repeats and hence it moves on to look for the signal "stop." The check started at clock cycle 19 and failed at clock cycle 23.

1.21.2 Consecutive repetition operator [*] on a sequence

Property p22 checks that two clock cycles after a valid start, sequence (a ##2 b) repeats 3 times and two clock cycles after that, signal "stop" is high.

```
Property p22;
 @(posedge clk) $rose(start) |->
          ##2 ((a ##2 b)[*3]) ##2 stop;
endproperty

a22: assert property(p22);
```

Figure 1-24 shows how property p22 reacts in a simulation. It shows 2 failures and one real success.

Failure 1 – The first failure is shown by marker 1s. A valid start is detected at this point. After 2 clock cycles from this point, the checker expects that the sequence (a ##2 b) repeats three times. But in this case, the sequence is repeated only 2 times. Hence, the checker fails and the failing point is shown by marker 1e.

Success 1 – The only real success is shown by marker 2s. A valid start is detected at this point. After 2 clock cycles from this point, the checker expects that the sequence (a ##2 b) repeats three times. The sequence is repeated 3 times as expected. A valid stop is expected 2 clock cycles after the successful repetition of the sequence and it happens as expected. Hence, the checker succeeds and the succeeding point is shown by marker 2e.

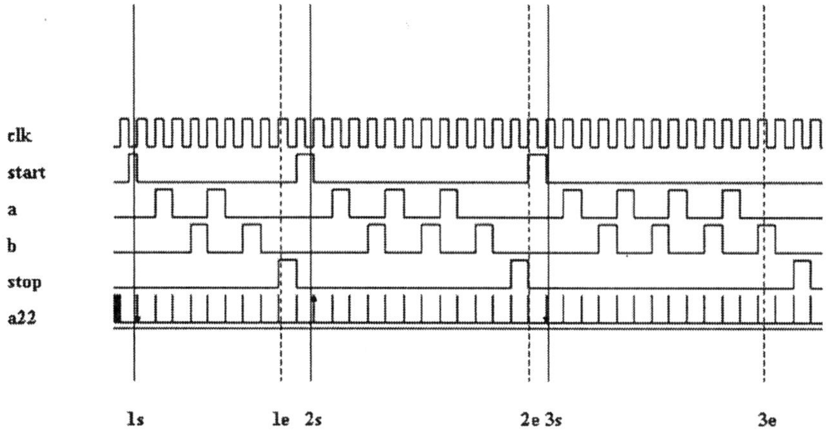

Figure 1-24. Waveform for SVA checker using consecutive repeat on a sequence

Failure 2 – The second failure is shown by marker 3s. A valid start is detected at this point. After 2 clock cycles from this point, the checker expects that the sequence (a ##2 b) repeats three times. The sequence repeats as expected. A valid stop is expected 2 clock cycles after the successful repetition of the sequence and it does not arrive. Hence, the checker fails and the failing point is shown by marker 3e.

1.21.3 Consecutive repetition operator [*] on a sequence with a delay window

Property p23 checks that two clock cycles after a valid start, sequence (a ##[1:4] b) repeats 3 times and two clock cycles after that, signal "stop" is high. The fact that the sequence has a timing window makes this check slightly complicated.

```
property p23;
  @(posedge clk) $rose(start) |->
      ##2 ((a ##[1:4] b)[*3]) ##2 stop;
endproperty

a23: assert property(p23);
```

The main sequence (a ##[1:4] b) [*3] expands as follows.

((a ##1 b) or (a ##2 b) or (a ##3 b) or (a ##4 b)) ##1
((a ##1 b) or (a ##2 b) or (a ##3 b) or (a ##4 b)) ##1
((a ##1 b) or (a ##2 b) or (a ##3 b) or (a ##4 b))

Figure 1-25 shows how property p23 reacts in a simulation. It shows 2 failures and one real success.

Failure 1 – The first failure is shown by marker 1s. A valid start is detected at this point. After 2 clock cycles from this point, the checker expects that the sequence (a ##[1:4] b) repeats three times. But in this case, the sequence is repeated only 2 times. Hence, the checker fails and the failing point is shown by marker 1e. Note that the 2 repeats that matched are (a ##1 b) and (a ##2 b) respectively.

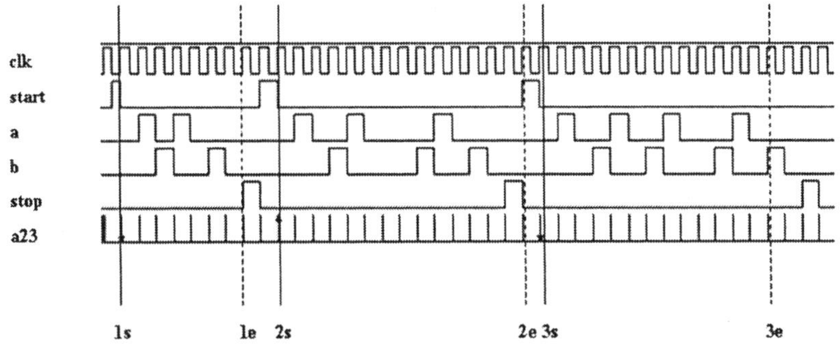

Figure 1-25. Waveform for SVA checker using consecutive repeat on a sequence with window of delay

Success 1 – The only real success is shown by marker 2s. A valid start is detected at this point. After 2 clock cycles from this point, the checker expects that the sequence (a ##[1:4] b) repeats three times. The sequence is repeated 3 times as expected. A valid stop is expected 2 clock cycles after the successful repetition of the sequence and it happens as expected. Hence, the checker succeeds and the succeeding point is shown by marker 2e. Note that the 3 repeats that matched are (a ##2 b), (a ##4 b) and (a ##2 b) respectively.

Failure 2 – The second failure is shown by marker 3s. A valid start is detected at this point. After 2 clock cycles from this point, the checker expects that the sequence (a ##[1:4] b) repeats three times. The sequence does repeat as expected. A valid stop is expected 2 clock cycles after the successful repetition of the sequence and it does not arrive as expected. Hence, the checker fails and the failing point is shown by marker 3e. Note that the 3 repeats that matched are (a ##2 b), (a ##2 b) and (a ##3 b) respectively.

1.21.4 Consecutive repetition operator [*] and eventuality operator

Property p23 specified a window of timing for the sequence that repeated itself. It is also possible to provide a window for the number of repetitions. For example, a [*1:5] means that signal "a" should repeat itself anywhere between 1 to 5 times. The definition can be expanded as follows.

```
a or
(a ##1 a) or
(a ##1 a ##1 a) or
```

```
(a ##1 a ##1 a ##1 a) or
(a ##1 a ##1 a ##1 a ##1 a)
```

The bounds of the repeat window follow the same rules as the delay windows. The left hand side value should be lesser than the right hand side value. The right hand side value can be a "$" sign indicating an unbounded number of repeats.

Property p24 shows an example of a finite check with an unbounded number of repeats defined. It checks that 2 cycles after a valid start signal, the signal "a" will stay high repeatedly until a valid stop arrives.

```
Property p24;
 @(posedge clk) $rose(start) |->
           ##2 (a[*1:$]) ##1 stop;
endproperty

a24: assert property(p24);
```

Figure 1-26 shows how property p24 reacts in a simulation. It shows one failure and one real success.

Failure 1 – A valid start occurs at clock cycle 3 shown by marker 1s. The check expects that 2 clock cycles from this point, signal "a" will stay high repeatedly until a valid stop arrives. Signal "a" detects high continuously until clock cycle 7. In clock cycle 8, it is detected low but the signal "stop" has not arrived yet. Hence, the check fails at clock cycle 8 shown by marker 1e.

Success 1 – A valid start occurs at clock cycle 11 shown by marker 2s. The check expects that 2 clock cycles from this point, signal "a" will stay high repeatedly until a valid stop arrives. Signal "a" stays high continuously until clock cycle 15. In clock cycle 16, it is detected low and the signal "stop" arrives as expected. Hence, the check succeeds at clock cycle 16 shown by marker 2e.

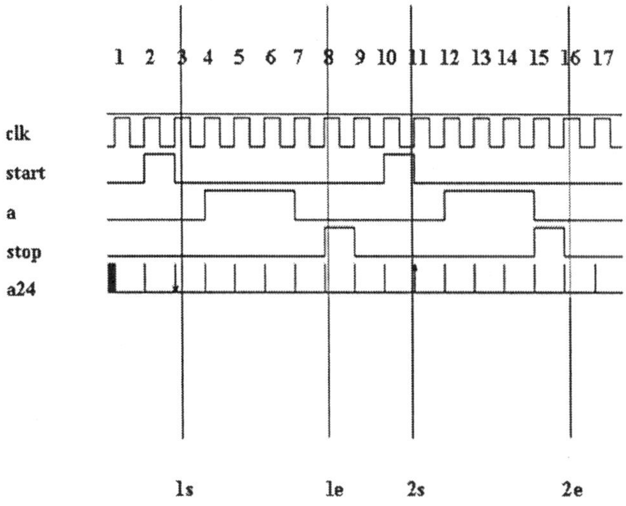

Figure 1-26. Waveform for SVA checker using consecutive repeat and eventuality

1.21.5 Go to repetition operator [->]

Property p25 checks that, if there is a valid start signal on any given positive edge of the clock, 2 clock cycles later, signal "a" will repeat three times continuously or intermittently before there is a valid stop signal.

```
property p25;
  @(posedge clk) $rose(start) |->
            ##2 (a[->3]) ##1 stop;
endproperty

a25: assert property(p25);
```

Figure 1-27 shows how property p25 reacts in a simulation. The figure shows that there is one failure, one real success and one incomplete check.

Failure 1 - A valid start of the checker is shown by marker 1s. The check expects that 2 clock cycles after the valid start, signal "a" will repeat three times. Signal "a" repeats 3 times as expected. After the third match on signal "a," a valid "stop" signal is expected on the next clock cycle. This does not happen and hence the check fails as shown by marker 1e.

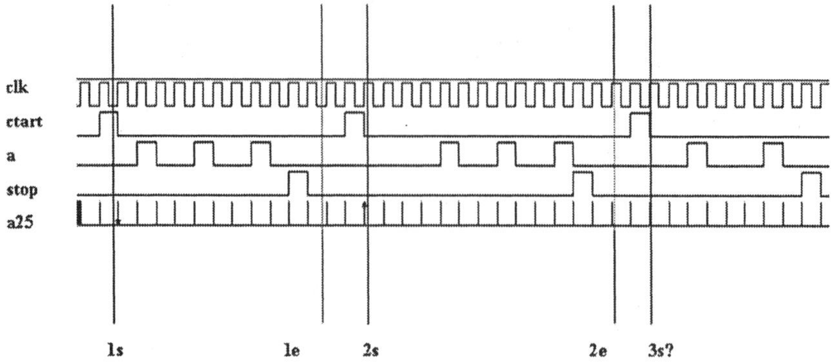

Figure 1-27. Waveform for SVA checker using go to repetition operator

Success 1 - A valid start of the checker is shown by marker 2s. The check expects that 2 clock cycles after the valid start, signal "a" will repeat three times. Signal "a" repeats 3 times as expected. After the third match on signal "a," a valid "stop" signal is expected on the next clock cycle. The "stop" signal arrives as expected and hence the check succeeds at marker 2e.

Incomplete 1 - A valid start of the checker is shown by marker 3s. The check expects that 2 clock cycles after the valid start, signal "a" will repeat three times. Signal "a" repeats 2 times; before the third one arrives, the simulation is finished. Also note that a valid "stop" signal arrives before the end of the simulation cycles. This "stop" will not have any effect since the repeat statement has not completed. The 3 expected repeats act as a blocking statement before the "stop" signal. Hence, the check is incomplete at the end of the simulation.

1.21.6 Non-consecutive repetition operator [=]

Property p26 checks that if there is a valid start signal on any given positive edge of the clock, 2 clock cycles later, signal "a" will repeat three times continuously or intermittently before there is a valid stop signal. One clock cycle later, the signal "stop" should be detected low. It checks for the exact same thing as property p25 except that it uses a "non-consecutive" repeat operator in the place of a "go to" repeat operator. This means that, in property p26, there is no expectation that there is a valid match on signal "a" in the previous cycle of a valid match on "stop" signal.

```
Property p26;
  @(posedge clk) $rose(start) |->
```

```
                  ##2 (a[=3]) ##1 stop ##1 !stop;
endproperty

a26: assert property(p26);
```

Figure 1-28 shows how property p26 reacts in a simulation. The figure shows that there are two successes and one incomplete check.

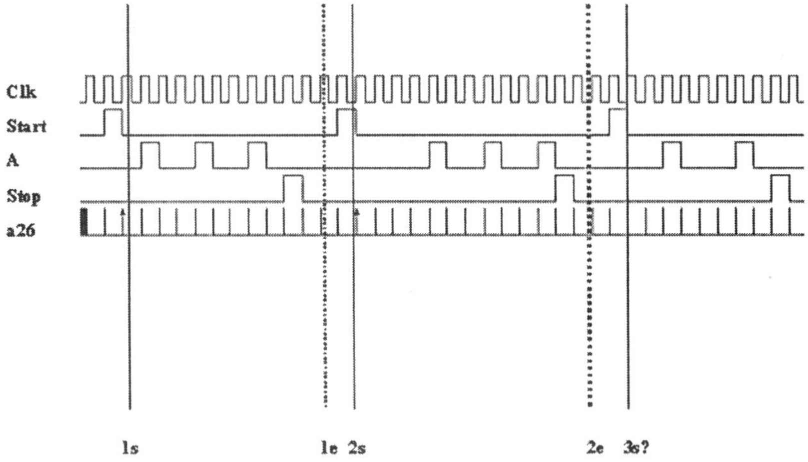

Figure 1-28. Waveform for SVA checker using non-consecutive repetition operator

Success 1 - A valid start of the checker is shown by marker 1s. The check expects that 2 clock cycles after the valid start, signal "a" will repeat three times. Signal "a" repeats 3 times as expected. After the third match on signal "a," a valid "stop" signal is expected, not necessarily on the next clock cycle. A valid "stop" signal arrives 2 clock cycles after the third match on signal "a" and hence the check succeeds as shown by marker 1e. This is the main difference between "go to" repetition and "non-consecutive" repetition. Property p25 failed for the same condition since a "go to" repetition was used.

Success 2 - A valid start of the checker is shown by marker 2s. The check expects that 2 clock cycles after the valid start, signal "a" will repeat three times. Signal "a" repeats 3 times as expected. After the third match on signal "a," a valid "stop" signal is expected, not necessarily on the next clock cycle. A valid "stop" signal arrives 1 clock cycle after the third match on signal "a" and hence the check succeeds as shown by marker 2e.

Incomplete 1 - A valid start of the checker is shown by marker 3s. The check expects that 2 clock cycles after the valid start, signal "a" will repeat three times. Signal "a" repeats 2 times; before the third one arrives, the simulation is finished. Also note that a valid "stop" signal arrives before the end of the simulation cycles. This "stop" will not have any effect since the repeat statement has not completed. The 3 expected repeats act as a blocking statement before the "stop" signal. Hence, the check is incomplete at the end of the simulation. This behavior is the same as in "go to" repetition.

1.22 The "and" construct

The binary operator "**and**" can be used to combine two sequences logically. The final property succeeds when both the sequences succeed. ***Both sequences must have the same starting point but they can have different ending points.*** The starting point of the check is when the first sequence succeeds and the end point is when the other sequence succeeds, ultimately making the property succeed.

Sequence s27a and s27b are two independent sequences. The property p27 combines them with an **and** operator. The property succeeds when both the sequences succeed.

```
sequence s27a;
  @(posedge clk) a##[1:2] b;
endsequence

sequence s27b;
  @(posedge clk) c##[2:3] d;
endsequence

property p27;
  @(posedge clk)  s27a and s27b;
endproperty

a27: assert property(p27);
```

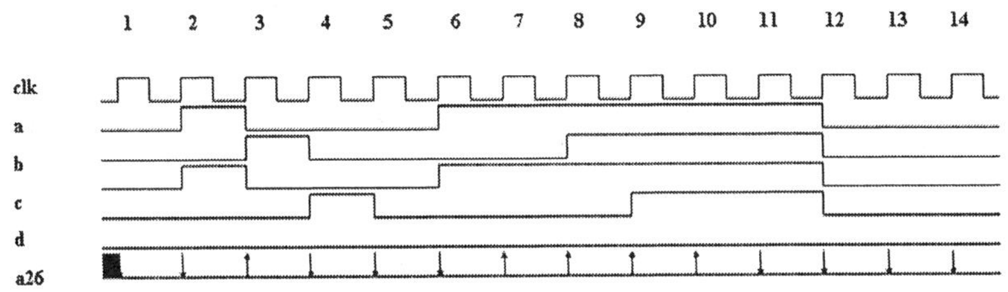

Figure 1-29. Waveform for SVA checker using "and" construct

Figure 1-29 shows how property p27 reacts in a simulation. Table 1-13 summarizes the sampled values of all relevant signals and the status of the assertion a27. There are 3 types of results. There can be a failure due to lack of a valid start. This happens on a given clock edge if signal "a" is not high or signal "c" is not high (clock cycles 1, 2, 4, 5, 6, 13, 14).

Table 1-13. Evaluation table for SVA checker using "and" construct

Clock cycle	Sampled value of "a"	Sampled value of "b"	Sampled value of "c"	Sampled value of "d"	Valid start	a27 status
1	0	0	0	0	No	Fail
2	0	0	0	0	No	Fail
3	1	0	1	0	Yes	Success (start at 3, end at 5)
4	0	1	0	0	No	Fail
5	0	0	0	1	No	Fail
6	0	0	0	0	No	Fail
7	1	0	1	0	Yes	Success (start at 7, end at 10)
8	1	0	1	0	Yes	Success (start at 8, end at 10)
9	1	1	1	0	Yes	Success (start at 9, end at 11)
10	1	1	1	1	Yes	Success (start at 10, end at 12)
11	1	1	1	1	Yes	Fail (start at 11, end at 14)
12	1	1	1	1	Yes	Fail (start at 12, end at 14)

Clock cycle	Sampled value of "a"	Sampled value of "b"	Sampled value of "c"	Sampled value of "d"	Valid start	a27 status
13	0	0	0	0	No	Fail
14	0	0	0	0	No	Fail

There are 5 different successes and each one of them has a different length. The valid checks that started at clock cycle 7 and clock cycle 8 both finish at clock cycle 10. For the check that starts at clock cycle 7, signal "b" is true in clock cycle 9, and signal "d" is true in clock cycle 10. For the check that starts at clock cycle 8, signal "b" is true in clock cycle 9, and signal "d" is true in clock cycle 10.

There are two failures, one at clock cycle 11 and one at clock cycle 12. Each one of them has the same length but they fail due to different reasons. For the check that starts at clock cycle 11, signal "b" is true in clock cycle 12. But signal "d" is never true in clock cycles 13 or 14 and hence the check fails at clock cycle 14. For the check that starts at clock cycle 12, signal "b" is not true in clock cycle 13. Both signal "b" and signal "d" are not true in clock cycle 14 and hence the check fails in clock cycle 14.

1.23 The "intersect" construct

The "**intersect**" operator is very similar to the "**and**" operator with one additional requirement. ***Both the sequences need to start at the same time and complete at the same time. In other words, the length of both sequences should be the same.***

Property p28 checks for the same condition as property p27. The only difference is that it uses the **intersect** construct instead of the **and** construct.

```
sequence s28a;
  @(posedge clk) a##[1:2] b;
endsequence

sequence s28b;
  @(posedge clk) c##[2:3] d;
endsequence

property p28;
  @(posedge clk)  s28a intersect s28b;
endproperty
```

```
a28: assert property(p28);
```

Figure 1-30 shows how property p28 reacts in a simulation. Table 1-14 summarizes the sampled values of all the relevant signals and the status of the assertion a28. Figure 1-30 also shows the results of assertion a27 that uses the **and** construct on the same set of design conditions. This helps understand the differences between the **and** construct and the **intersect** construct.

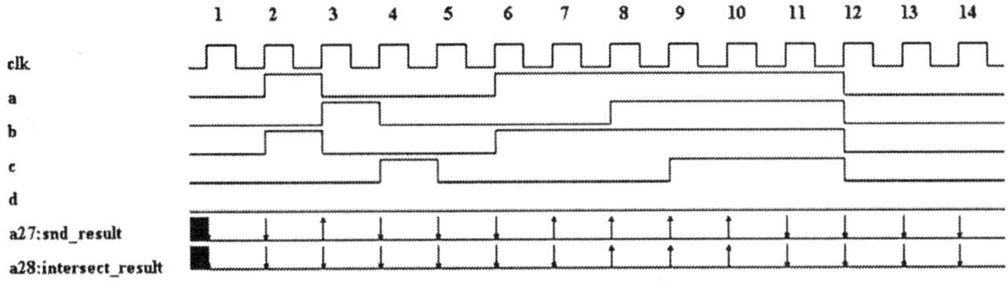

Figure 1-30. Waveform for SVA checker using "intersect" construct

The failures due to lack of a valid start remain the same. The second set of failure happens from the fact that the individual sequences do not match as expected. This kind of failure happens at clock cycles 11 and 12. The third set of failure happens even though the individual sequences match as expected. ***These failures happen since the individual sequence did not take the same length of time to match.*** In the failure shown in clock cycle 3, sequence s28a takes one clock cycle to match ("a" is true in clock cycle 3 and "b" is true in clock cycle 4) and sequence s28b takes 2 clock cycles to match ("c" is true in clock cycle 3 and "d" is true in clock cycle 5). In the failure shown in clock cycle 7, s28a takes two clock cycles to match ("a" is true in clock cycle 7 and "b" is true in clock cycle 9) and sequence s28b takes three clock cycles to match ("c" is true in clock cycle 7 and "d" is true in clock cycle 10).

The three successes happen at clock cycles 8, 9 and 10 respectively. In all these three cases the sequences match with the same length of time.

In the success shown in clock cycle 8, sequence s28a matches twice, at clock cycles 9 and 10. The sequence s28b also matches twice, at clock cycles 10 and 11. The common length is 2 clock cycles for both the sequences to match. Hence, the **intersect** succeeds with s28a matching at clock cycle 10 and s28b also matching at clock cycle 10. Each of them has a length of 2 clock cycles.

Table 1-14. Evaluation table for SVA checker using "intersect" construct

Clock tick	Sampled value of "a"	Sampled value of "b"	Sampled value of "c"	Sampled value of "d"	Valid start	a28 status
1	0	0	0	0	No	Fail
2	0	0	0	0	No	Fail
3	1	0	1	0	Yes	**Fail (sequences succeed with different length)**
4	0	1	0	0	No	Fail
5	0	0	0	1	No	Fail
6	0	0	0	0	No	Fail
7	1	0	1	0	Yes	**Fail (sequences succeed with different length)**
8	1	0	1	0	Yes	Success (start at 8, end at 10)
9	1	1	1	0	Yes	Success (start at 9, end at 11)
10	1	1	1	1	Yes	Success (start at 10, end at 12)
11	1	1	1	1	Yes	Fail (start at 11, end at 13)
12	1	1	1	1	Yes	Fail (start at 12, end at 14)
13	0	0	0	0	No	Fail
14	0	0	0	0	No	Fail

In the success shown in clock cycle 9, sequence s28a matches twice, at clock cycles 10 and 11. The sequence s28b also matches twice, at clock cycles 11 and 12. The common length is 2 clock cycles for both the sequences to match. Hence, the **intersect** succeeds with s28a matching at clock cycle 11 and s28b also matching at clock cycle 11. Each of them has a length of 2 clock cycles.

In the success shown in clock cycle 10, sequence s28a matches twice, at clock cycles 11 and 12. The sequence s28b matches at clock cycle 12. The common length is 2 clock cycles for both the sequences to match. Hence, the

intersect succeeds with s28a matching at clock cycle 12 and s28b also matching at clock cycle 12. Each of them has a length of 2 clock cycles.

1.24 The "or" construct

The binary operator "**or**" can be used to combine two sequences logically. ***The final property succeeds when any one of the sequence succeeds.***

Sequence s29a and s29b are two independent sequences. The property p29 combines them with an **or** operator. The property succeeds when any one of the sequence succeeds.

```
sequence s29a;
 @(posedge clk) a##[1:2] b;
endsequence

sequence s29b;
 @(posedge clk) c##[2:3] d;
endsequence

property p29;
 @(posedge clk)  s28a or s28b;
endproperty

a29: assert property(p29);
```

Figure 1-31 shows how property p29 reacts in a simulation. Table 1-15 summarizes the sampled values of all the relevant signals and the status of the assertion a29. Figure 1-31 also shows the results of assertion a27 that uses the **and** construct on the same set of design conditions. This helps understand the differences between the **and** construct and the **or** construct. The failures due to the lack of a valid start remain the same. The second set of failure happens from the fact that the individual sequences do not match as expected. This kind of failure happens at clock cycle 12. Both sequences never match within their timing window and hence the check fails.

The successes are almost the same for the **and** operator and **or** operator. The main difference is the duration of the match. The **or** operator matches as soon as a match is found on sequence s29a and hence does not wait for sequence s29b to finish.

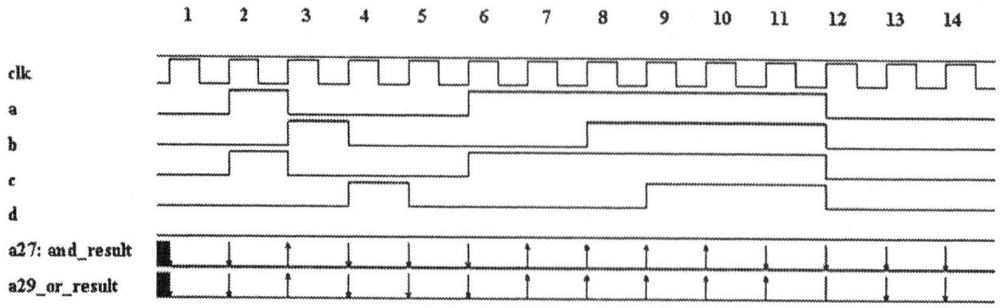

Figure 1-31. Waveform for SVA checker using "or" construct

Table 1-15. Evaluation table for SVA checker using "or" construct

Clock tick	Sampled value of "a"	Sampled value of "b"	Sampled value of "c"	Sampled value of "d"	Valid start	a29 status
1	0	0	0	0	No	Fail
2	0	0	0	0	No	Fail
3	1	0	1	0	Yes	Success (start at 3, end at 4)
4	0	1	0	0	No	Fail
5	0	0	0	1	No	Fail
6	0	0	0	0	No	Fail
7	1	0	1	0	Yes	Success (start at 7, end at 9)
8	1	0	1	0	Yes	Success (start at 8, end at 9)
9	1	1	1	0	Yes	Success (start at 9, end at 10)
10	1	1	1	1	Yes	Success (start at 10, end at 11)
11	1	1	1	1	Yes	Success (start at 11, end at 12)
12	1	1	1	1	Yes	Fail (start at 12, end at 14)
13	0	0	0	0	No	Fail
14	0	0	0	0	No	Fail

One of the failures with the **and** construct at clock cycle 11 becomes a success with the **or** construct. The reason for this is that the first part of sequence s29a matches at clock cycle 12 and this immediately makes the property succeed. In the **and** construct this alone is not enough. The second part of the sequence has to match, but it does not occur within the specified time window. Therefore, the same condition makes the property p27 fail at clock cycle 14.

1.25 The "first_match" construct

Whenever a timing window is specified in sequences along with binary operators such as **and** and **or,** there is a possibility of getting multiple matches for the same check. The construct "**first_match**" ensures that only the first sequence match is used and the others are discarded. This becomes very helpful when combining multiple sequences together wherein only the first match in the timing window is required to evaluate the remaining part of the property.

In the example shown below, two sequences are combined with an **or** operator. There are several possible matches for this property and they are as follows.

```
a ##1 b;
a ##2 b;
c ##2 d;
a ##3 b;
c ##3 d;
```

When the property p30 gets evaluated, the first one to match will be kept and every other match will be discarded.

```
sequence s30a;
 @(posedge clk) a ##[1:3] b;
endsequence

sequence s30b;
 @(posedge clk) c ##[2:3] d;
endsequence

property p30;
 @(posedge clk) first_match(s30a or s30b);
endproperty
```

```
a30: assert property(p30);
```

Figure 1-32 shows how property p30 reacts in a simulation. There are 2 successes shown in the figure, one at clock cycle 3 and another at clock cycle 9. The success at clock cycle 3 is based on the match on the sequence (c ##2 d). The success at clock cycle 9 is based on the match on the sequence (a ##1 b). In both cases, the first sequence match made the property succeed.

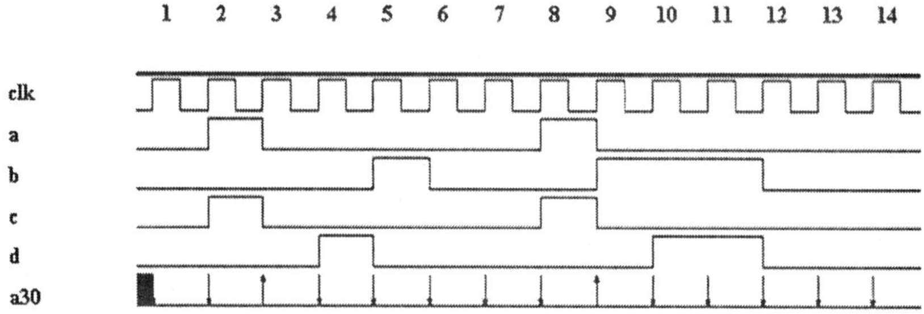

Figure 1-32. Waveform for SVA checker using "first_match" construct

1.26 The "throughout" construct

Implication is one technique discussed so far that allows defining pre-conditions. For example, for a specific sequence to be tested, a certain pre-condition must be true. There are also situations wherein the condition must hold true until the entire test sequence completes. Implication checks for pre-condition once on the clock edge and then starts evaluating the consequent part. Hence, it does not care if the antecedent remains true or not. To make sure that certain condition holds true during the evaluation of the entire sequence, "**throughout**" operator should be used. The simple syntax of a **throughout** operator is shown below.

```
(expression) throughout (sequence definition)
```

Property p31 checks the following.

a. The check starts when signal "start" has a falling edge.
b. Test the expression ((!a&&!b) ##1 (c[->3]) ##1 (a&&b)).
c. The sequence checks that between the falling edge of signals "a"

and "b," and the rising edge of signals "a" and "b," signal "c" should repeat itself 3 times continuously or intermittently.

d. During the entire test expression, signal "start" should always be low.

```
property p31;
  @(posedge clk) $fell(start) |->
        (!start) throughout
        (##1 (!a&&!b) ##1 (c[->3]) ##1 (a&&b));
endproperty

a31: assert property(p31);
```

Figure 1-33 shows how property p31 reacts in a simulation. The check succeeds at clock cycle 3 and fails in clock cycle 16.

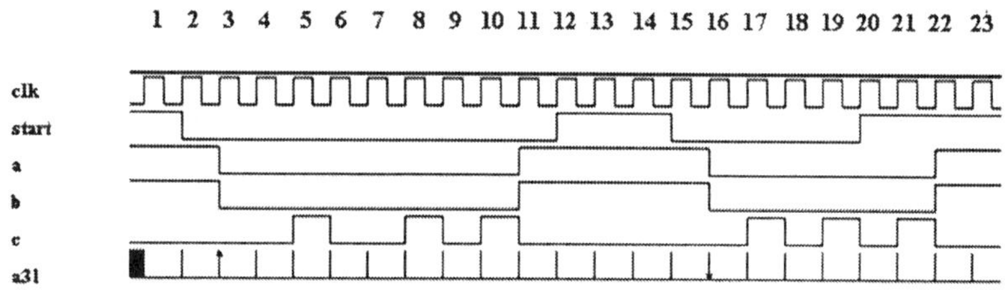

Figure 1-33. Waveform for SVA checker using "throughout" construct

Success 1 – The antecedent of the property succeeds on clock cycle 3 when a falling edge is detected on the start "signal." One cycle after that, signals "a" and "b" are expected to be low, and they are as expected in clock cycle 4. From this point, signal "c" is expected to repeat itself three times. It does repeat three times, once each in clock cycles 6, 9 and 11. In clock cycle 12, it is expected that both signals "a" and "b" are high, and they are as expected. Hence, the property starts at clock 3 and succeeds at clock 12. ***Note that signal "start" was detected low from the clock cycles 3 through 12. That is the key for the success of this check.***

Failure 1 – The antecedent of the property succeeds on clock cycle 16 when a falling edge is detected on the "start" signal. One cycle after that, signals "a" and "b" are expected to be low, and they are in clock cycle 17. From this point signal, "c" is expected to repeat itself three times. We get two repeats on clock cycles 18 and 20. But on clock 21, before the third repeat on signal "c" arrives, the signal "start" is detected high and the check fails at clock cycle 21. The "throughout" condition was violated here and hence the check fails.

1.27 The "within" construct

The "**within**" construct allows the definition of a sequence contained within another sequence.

```
seq1 within seq2
```

This means that seq1 happens within the start and completion of seq2. The starting matching point of seq2 must happen before the starting matching point of seq1. The ending matching point of seq1 must happen before the ending matching point of seq2. Property p32 checks that the sequence s32a happens within the rise and fall of signal "start." The rise and fall of signal "start" is defined as a sequence in s32b.

```
sequence s32a;
 @(posedge clk)
       ((!a&&!b) ##1 (c[->3]) ##1 (a&&b));
endsequence

sequence s32b;
 @(posedge clk)
       $fell(start) ##[5:10] $rose(start);
endsequence

sequence s32;
 @(posedge clk) s32a within s32b;
endsequence

property p32;
 @(posedge clk) $fell(start) |-> s32;
endproperty

a32: assert property(p32);
```

The same set of design conditions used to describe the **throughout** operator is used in Figure 1-34 to show how property p32 reacts in a simulation. There are two valid starts for this check, one at clock cycle 3 and another at clock cycle 16. In both these clocks, a falling edge of the signal "start" is detected.

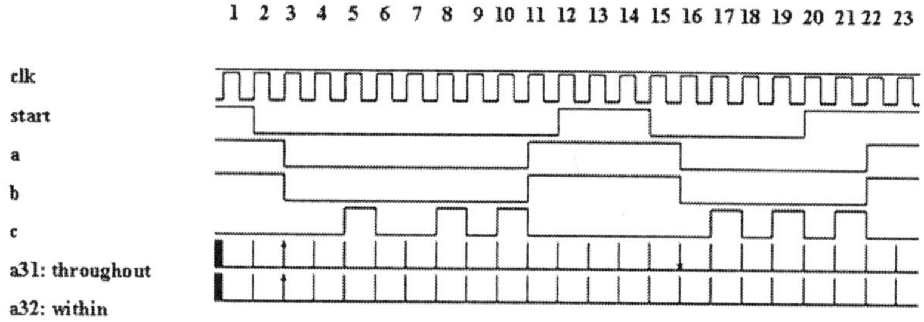

Figure 1-34. Waveform for SVA checker using "within" construct

Success 1 – The check starting at clock cycle 3 succeeds. The falling edge of signal "start" is at clock cycle 3 and the rising edge of the signal "start" is at clock cycle 13. Within these clock cycles, signal "c" is detected high three times in clock cycles 6, 9 and 11. Hence, the check succeeds.

Incomplete 1 - The check starting at clock cycle 16 never finishes. The falling edge of signal "start" is at clock cycle 16 and the rising edge of the signal "start" is at clock cycle 21. Within these clock cycles, signal "c" is detected high two times in clock cycles 18 and 20 respectively. The third repeat of signal "c" comes at clock cycle 22 but signal "start" is detected high at clock cycle 21. This is a failure but since a "go to" repetition operator is used to check for signal "c," it acts as a blocking sequence. This makes the check fail and issues an incomplete message during simulation.

1.28 Built-in system functions

SVA provides several built-in functions to check for some of the most common design conditions.

$onehot(expression) – checks that the expression is one-hot, in other words, only one bit of the expression can be high on any given clock edge.

$onehot0(expression) – checks that the expression is zero one-hot, in other words, only one bit of the expression can be high or none of the bits can be high on any given clock edge.

$isunknown(expression) – checks if any bit of the expression is X or Z.

$countones(expression) – counts the number of bits that are high in a vector.

Assert statement a33a checks that the bit vector "state" is one-hot. Assert statement a33b checks that the bit vector "state" is zero one-hot. Assert statement a33c checks if any bit of the vector "bus" is X or Z. Assert statement a33d checks that the number of ones in the vector "bus" is greater than one.

```
a33a: assert
      property(@(posedge clk) $onehot(state));
a33b: assert
      property(@(posedge clk) $onehot0(state));
a33c: assert
      property(@(posedge clk) $isunknown(bus));
a33d: assert
      property(@(posedge clk)$countones(bus)> 1);
```

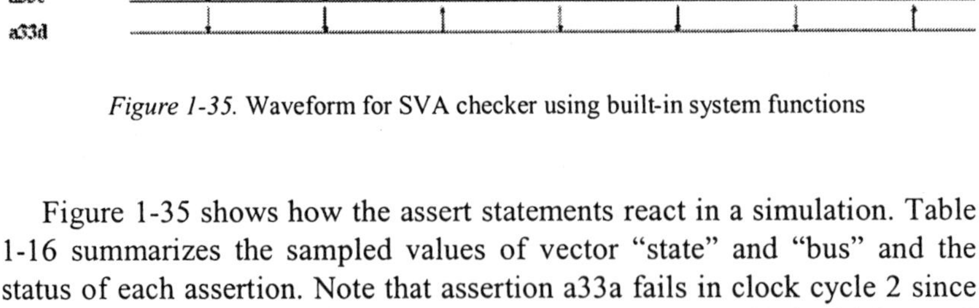

Figure 1-35. Waveform for SVA checker using built-in system functions

Figure 1-35 shows how the assert statements react in a simulation. Table 1-16 summarizes the sampled values of vector "state" and "bus" and the status of each assertion. Note that assertion a33a fails in clock cycle 2 since all bits are zero. The one-hot condition requires that one bit be high on all positive edges of the clock. On the other hand, assertion a33b passes since it checks for zero one-hot and all bits being zero is legal for this construct. Both a33a and a33b fail in clock cycles 5, 6, 7 and 8 wherein more than one bit is high. Assertion a33c fails anytime the value of the vector "bus" is not Z or X. It passes on clock cycles 5, 6 and 7 wherein the value is Z. Assertion a33d fails on clock cycles 2, 3, 5, 6, and 7 wherein no more than one bit is high. Assertion a33d passes in clock cycles 4 and 8 since 2 bits are high in the vector "bus" on these two clock cycles.

Table 1-16. Evaluation table for SVA checker using built-in functions

Clock tick	Sampled value of "state"	Sampled value of "bus"	a33a - $onehot status	a33b - $onehot0 status	a33c - $isunknown status	a33d - $countones status
2	0000	00100	Fail	Success	Fail	Fail
3	0010	00001	Success	Success	Fail	Fail
4	0100	01001	Success	Success	Fail	Success
5	0011	Z	Fail	Fail	Success	Fail
6	1101	Z	Fail	Fail	Success	Fail
7	1101	Z	Fail	Fail	Success	Fail
8	1101	01010	Fail	Fail	Fail	Success

1.29 The "disable iff" construct

In certain design conditions, we don't want to proceed with the check if some condition is true. In other words, it is like an asynchronous reset that will make the check currently being evaluated void. SVA provides a construct called "**disable iff**" that acts like an asynchronous reset for the checker. The simple syntax for a **disable iff** is as follows.

disable iff (expression) < property definition>

Property p34 checks that after a valid start, signal "a" repeat 2 times and 1 cycle after that, signal "b" repeats 2 times and one cycle later signal "start" becomes low. During this entire sequence, if reset is detected high at any point, the checker will stop and issue a vacuous success by default.

```
property p34;
   @(posedge clk)
  disable iff (reset)
  $rose(start) |=> a[=2] ##1 b[=2] ##1 !start ;
endproperty

a34: assert property(p34);
```

Figure 1-36 shows how property p34 reacts in a simulation. A valid start is shown with marker 1s. After the valid start, signal "a" repeats two times and then signal "b" repeats two times. Signal "start" becomes low after that as expected.

During this entire sequence, the signal "reset" is inactive as expected and hence the check succeeds at marker 1e. A second valid start is shown with marker 2s. After the valid start, signal "a" repeats two times and then the "reset" signal becomes active before signal "b" could repeat two times. This nullifies the check and the property succeeds vacuously.

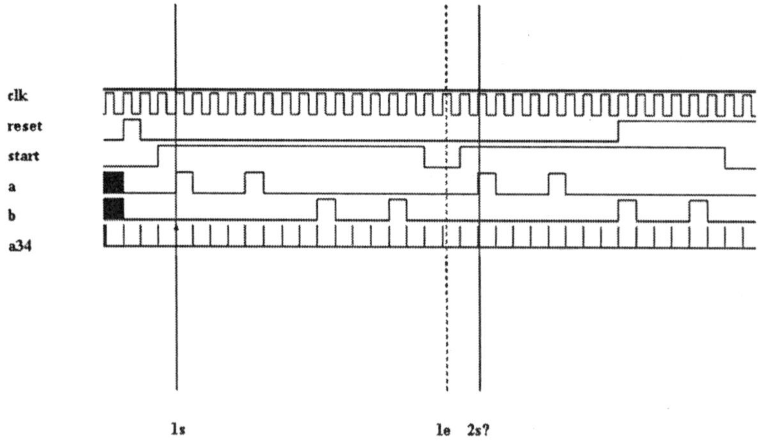

Figure 1-36. Waveform for SVA checker using "disable iff" construct

1.30 Using "intersect" to control length of the sequence

The **intersect** operator discussed in Section 1.23 can be used effectively to control the length of sequences, particularly in cases where the upper bound of the timing window is not defined. Whenever an eventuality operator is used, there is no restriction on the number of clock cycles that can be used by the checker to succeed. The **intersect** operator provides a mechanism to define the minimum and maximum number of clock cycles that can be used by the eventuality operator to succeed.

Property p35 defines a sequence that checks that on a given clock edge if signal "a" is high then eventually signal "b" should go high starting from the next clock cycle and eventually signal "c" should go high starting from the next clock cycle. This sequence will start whenever signal "a" is high and can take until the end of the simulation time to succeed. This is restricted by using the **intersect** operator 1[*2:5]. This **intersect** definition checks that from the starting point of the sequence match (high on signal "a") to the ending point of the sequence match (high on signal "c") it can take anywhere between 2 to 5 clock cycles.

```
Property p35;
   (@(posedge clk)  1[*3:5] intersect
                    (a ##[1:$] b ##[1:$] c));
endproperty

a35: assert property(p35);
```

Figure 1-37 shows how property p35 reacts in a simulation. Table 1-17 summarizes the sampled values of the relevant signals and shows the status of assertion a35. On a given clock edge if signal "a" is not detected high, it is a failure. This happens in several clock cycles (1, 3, 4, 5, 11 and 13) and these are not valid starts.

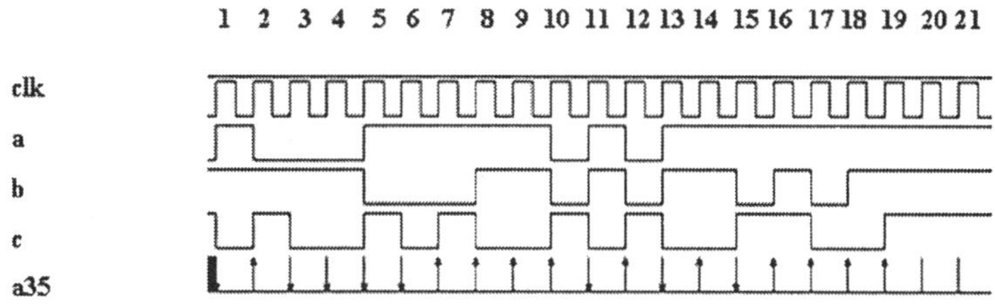

Figure 1-37. Waveform for SVA checker using intersect to control the length of the sequence

The check succeeds in several clock cycles (2, 6, 7, 8, 9, 10, 12 and 14). Note that the sequence takes 5 clock cycles or less from the start to the end point. The check has a real failure at clock cycle 6. Signal "a" is detected high at clock cycle 6 and signal "b" arrives at clock cycle 9. Signal "c" does not arrive at clock cycle 10, which completes the upper limit allowed for the length of the entire check. Hence, the check fails at clock 10. Note that signal "c" does arrive at clock cycle 11, but it's too late.

Table 1-17. Evaluation table for SVA checker using intersect operator to control the length of the sequence

Clock tick	Sampled value of "a"	Sampled value of "b"	Sampled value of "c"	Valid start	a35 status
1	0	1	1	No	Fail
2	1	1	0	Yes	Success (start at 2, end at 6)
3	0	1	1	No	Fail
4	0	1	0	No	Fail
5	0	1	0	No	Fail
6	1	0	1	Yes	Fail (start at 6, end at 10)
7	1	0	0	Yes	Success (start at 7, end at 11)
8	1	0	1	Yes	Success (start at 8, end at 11)
9	1	1	0	Yes	Success (start at 9, end at 11)
10	1	1	0	Yes	Success (start at 10, end at 13)
11	0	0	1	No	Fail
12	1	1	0	Yes	Success (start at 12, end at 16)
13	0	0	1	No	Fail
14	1	1	0	Yes	Success (start at 14, end at 16)
15	1	1	0	Yes	Fail
16	1	0	1	Yes	Success
17	1	1	1	Yes	Success

1.31 Using formal arguments in a property

Some of the common properties that can be re-used can be defined with formal arguments. Property "arb" takes 4 formal arguments and has a check defined on these formal arguments. The property is also bound to a specific clock. SVA allows clock definition as one of the formal arguments to the property. This way, the property can be bound to similar design block working with different clocks. Also, the timing delays specified can be parameterized to make the property definition very generic.

The property checks for a valid start first. On a given positive edge of the clock, if a falling edge of signal "a" is followed by the falling edge of signal "b" within 2 to 5 clock cycles, then it is a valid start.. If the antecedent matches, then the property checks for a falling edge on signal "c" and signal "d" on the next clock cycle and makes sure that these two signals stay low for 4 consecutive cycles. One cycle later, signal "c" and signal "d" should be detected high and one cycle after that signal b should be detected high.

Assuming that this is a protocol followed by an arbiter that deals with three different master devices with similar signals, the property can be re-used easily to check all three master interfaces. Assertions a36_1, a36_2 and a36_3 define the assertions for each master interface, using the signals relevant to each interface as the arguments for the property.

```
property arb (a, b, c, d);
  @(posedge clk) ($fell(a) ##[2:5] $fell(b)) |->
     ##1 ($fell(c) && $fell(d)) ##0
     (!c&&!d) [*4] ##1 (c&&d) ##1 b;
endproperty

a36_1: assert property(arb(a1, b1, c1, d1));
a36_2: assert property(arb(a2, b2, c2, d2));
a36_3: assert property(arb(a3, b3, c3, d3));
```

Figure 1-38 shows how the assertions defined for each interface react in a simulation. Assertion a36_1 has one valid start and it succeeds. Assertion a36_3 has one valid start and it fails.

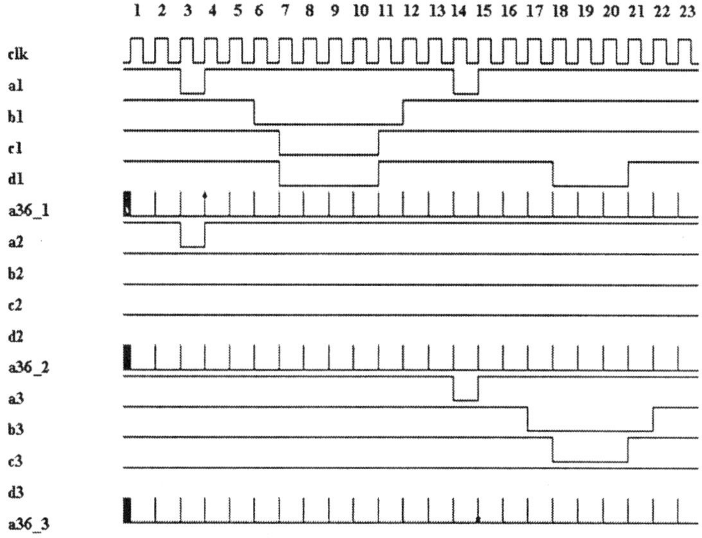

Figure 1-38. Waveform for SVA checker using formal arguments in a property

Success 1 (a36_1) - The check begins when a falling edge arrives on signal "a1" on clock cycle 4. This expects that a falling edge arrives on signal "b1" within 2 to 5 clock cycles and it does arrive on clock cycle 7. In the next clock cycle, signal "c1" and "d1" are low as expected. They should remain low for four cycles. They remain low from clock cycle 8 to 11. At clock cycle 12, both the signals "c1" and "d1" are high as expected. At clock cycle 13, signal "b1" is high as expected. Hence, the signal starts at clock cycle 4 and succeeds at clock cycle 13.

Failure 1 (a36_3) - The check begins when a falling edge arrives on signal "a3" on clock cycle 15. This expects that a falling edge arrives on signal "b3" within 2 to 5 clock cycles and it does arrive on clock cycle 18. In the next clock cycle, signal "c3" and "d3" are expected to be low. Since signal "d3" is not detected to be low, the check fails at clock cycle 19.

1.32 Nested implication

SVA allows having nested implications. These are useful when we have multiple gating conditions leading to a single final consequent.

Property p_nest checks that a valid start occurs if there is a falling edge on signal "a," then one cycle later, signals "b," "c" and "d" should all be active low to keep the valid start alive. If the second condition matches, then it is expected that within 6 to 10 cycles the condition "free" is true. Note that the consequent condition "true" is evaluated if and only if the signals "b," "c" and "d" match as expected.

```
`define free (a && b && c && d)

property p_nest;
   @(posedge clk) $fell(a) |->
                  ##1 (!b && !c && !d) |->
                                   ##[6:10] `free;
endproperty

a_nest: assert property(p_nest);
```

The same property can be re-written without using the nested implication as follows.

```
property p_nest1;
   @(posedge clk) $fell(a) ##1 (!b && !c && !d)
```

```
                         |-> ##[6:10] `free;
endproperty

a_nest1: assert property(p_nest1);
```

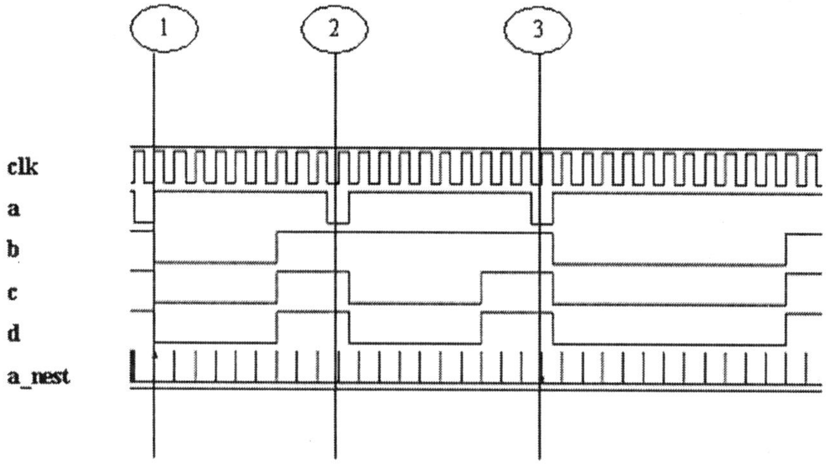

Figure 1-39. SVA checker with nested implication

Note that the nested implication property p_nest has no "else" condition and hence, the property can be easily re-written as shown in p_nest1.

Figure 1-39 shows how the assertion a_nest behaves in a simulation. Marker 1 shows the first success of the checker. A valid start occurs when a falling edge is detected on signal "a." One cycle later, signals "b," "c" and "d" are detected low as expected and hence the check is kept alive and the consequent gets evaluated. The condition "free" is detected true 6 clock cycles later and hence the check succeeds.

The second marker indicates the next valid start wherein a falling edge of signal "a" is detected. One cycle later, signals "c" and "d" are detected low but signal "b" is not low. Hence, the check is not active anymore and the check succeeds vacuously.

The third marker indicates a valid start wherein a falling edge of signal "a" is detected. One cycle later, signals "b," "c" and "d" are detected low as expected and hence the check is still active and the consequent gets evaluated. The condition "free" is not detected true within 6 to 10 clock cycles after and hence the check fails.

1.33 Using if/else with implication

SVA allows the use of an "**if/else**" statement on the consequent of an implied property. Property p_if_else checks that a valid start occurs if a falling edge is detected on signal "start" and one clock cycle later either signal "a" or signal "b" is detected high. On a successful match of the antecedent, the consequent can take two possible paths.

1. If signal "a" is detected high, then, signal "c" should repeat twice intermittently and one cycle later signal "e" should be high.
2. If signal "a" is not high, then, signal "d" should repeat twice intermittently and one cycle later signal "f" should be high.

Note that there is a priority in the evaluation of the consequent for signal "a."

```
property p_if_else;
  @(posedge clk)
  ($fell(start) ##1 (a||b)) |->
    if(a)
      (c[->2] ##1 e)
    else
      (d[->2] ##1 f);
endproperty

a_if_else: assert property(p_if_else);
```

To re-write this property without using an "if/else" construct, three separate properties are required. A priority based "if/else" on two signals leads to three different possibilities as shown below.

a	b	Leaf
1	0	a
0	1	b
1	1	a

Note that if both signals "a" and "b" are high, then the "if" block of signal "a" is executed since it has priority. The three properties are shown below.

```
property p_if_else_leaf1;
 @(posedge clk)
       ($fell(start) ##1 a) |->
                            (c[->2] ##1 e);
endproperty

a_if_else_leaf1:
      assert property(p_if_else_leaf1);

property p_if_else_leaf2;
 @(posedge clk)
       ($fell(start) ##1 b) |->
                            (d[->2] ##1 f);
endproperty

a_if_else_leaf2:
      assert property(p_if_else_leaf2);

property p_if_else_leaf3;
 @(posedge clk)
       ($fell(start) ##1 (a &&b)) |->
                            (c[->2] ##1 e);
endproperty

a_if_else_leaf3:
      assert property(p_if_else_leaf3);
```

1.34 Multiple clock definitions in SVA

SVA allows a sequence or a property to have multiple clock definitions for sampling individual signals or sub-sequences. SVA will automatically synchronize between the clock domains used in the signals or sub-sequences. The following code shows a simple example of a sequence using multiple clocks.

```
sequence s_multiple_clocks;
     @(posedge clk1) a ##1 @(posedge clk2) b;
```

```
endsequence
```

The sequence s_multiple_clocks checks that, on a given positive edge of clock "clk1," signal "a" is high and then on a give positive edge of clock "clk2," signal "b" is high. The sequence matches when signal "a" is high on any given positive edge of clock "clk1." The ##1 delay construct will move the evaluation time to the nearest positive clock edge of clock "clk2" and then will check for signal "b" being high. ***When multiple clocked signals are used in a sequence, only ##1 delay construct is allowed.*** Re-writing the sequence s_multiple_clocks as follows is not allowed.

```
sequence s_multiple_clocks_illegal1;
     @(posedge clk1) a ##0 @(posedge clk2) b;
endsequence

sequence s_multiple_clocks_illegal2;
     @(posedge clk1) a ##2 @(posedge clk2) b;
endsequence
```

The use of ##0 will create confusion on which one is the nearest clock after the match on signal "a." This will create race conditions; hence, it is not allowed. The use of ##2 is not allowed since it is not possible to synchronize to the nearest positive clock edge of clock "clk2."

Similar techniques can be used to create properties with multiple clocks. The following code shows an example.

```
property p_multiple_clocks;
 @(posedge clk1) s1 ##1 @(posedge clk2) s2;
endproperty
```

It is assumed that the sequence s1 is not clocked or it has the same clock definition as "clk1." It is assumed that the sequence s2 is not clocked or it has the same clock definition as "clk2." The property can also have a non-overlapping implication operator in between the sequence definitions. A sample code is shown below.

```
property p_multiple_clocks_implied;
 @(posedge clk1) s1 |=> @(posedge clk2) s2;
endproperty
```

The use of an overlapping implication operator between two multiple clocked sequences is not allowed. Since the end of the antecedent and the

beginning of the consequent overlaps, it can lead to race conditions; hence, it is illegal. The following code shows the illegal coding style.

```
property p_multiple_clocks_implied_illegal;
 @(posedge clk1) s1 |-> @(posedge clk2) s2;
endproperty
```

1.35 The "matched" construct

Whenever a sequence is defined with multiple clocks, the construct "**matched**" is used to detect the endpoint of the first sequence. Sequence s_a looks for a rising edge on the signal "a." Signal "a" is sampled based on the clock "clk1." Sequence s_b looks for a rising edge on the signal "b." Signal "b" is sampled based on the clock "clk2." The property p_match verifies that on a given positive edge of clock "clk2," if there is a match on sequence s_a, then one cycle later sequence s_b should be true.

```
sequence s_a;
 @(posedge clk1) $rose(a);
endsequence

sequence s_b;
 @(posedge clk2) $rose(b);
endsequence

property p_match;
 @(posedge clk2) s_a.matched |=> s_b;
endproperty

a_match: assert property(p_match);
```

Figure 1-40 shows how the assertion a_match behaves in a simulation. The property gets a valid start when there is a match on sequence s_a. Note that we are looking for this match on every positive edge of clock "clk2," though sequence s_a is sampled based on clock "clk1."

A valid rise on signal "a" happens at clock cycle 3 of "clk1." This updates the match value on sequence s_a to true. This value will be held until the nearest positive clock edge of "clk2." The nearest positive edge of "clk2" is at clock cycle 2 of "clk2." At this point the property becomes active and one clock cycle of "clk2" later, it is expected that the sequence s_b matches. Hence, the first success of the property starts at clock cycle 2 of "clk2" and ends at clock cycle 3 of "clk2."

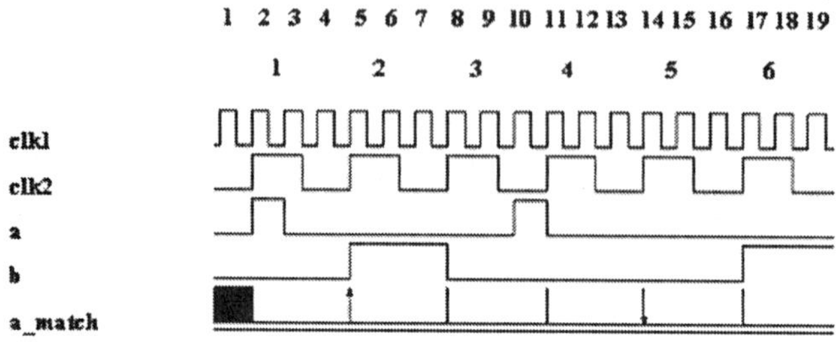

Figure 1-40. SVA checker using "matched" construct

Another valid rise on signal "a" happens at clock cycle 11 of "clk2" and this is sampled by the property at clock cycle 5 of "clk2." The property becomes active at this point and it is expected that in clock cycle 6 of "clk2," the sequence s_b match. But in this case, a rising edge of signal "b" does not occur and hence the property fails. ***The key concept to understand in using "matched" construct is that, the sampled match value is stored only until the next nearest clock edge of the other sequence.***

1.36 The "expect" construct

SVA supports a construct called "**expect,**" which is similar to the wait construct in Verilog, with the key difference being that the expect statement waits on the successful evaluation of a property. It acts as a blocking statement for the code that follows the **expect** construct. The syntax of the **expect** construct is very similar to the **assert** construct. ***The expect statement is allowed to have an action block upon the success or failure of the property.*** A sample code using the **expect** construct is shown below.

```
initial

begin
@(posedge clk);
#2ns cpu_ready = 1'b1;
expect(@(posedge clk) ##[1:16]
                    memory_ready == 1'b1)
```

```
    $display("Hand shake successful\n");
  else
    begin
    $display("Hand shake failed: exiting\n")
    $finish();
    end

  for(i=0; i<64; i++)
  begin
    send_packet();
    $display("PACKET %0d sent\n", i);
  end

  end
```

Note that after the signal "cpu_ready" is asserted, the **expect** statement waits for anywhere between 1 to 16 cycles for the signal "memory_ready" to be asserted. If the signal "memory_ready" is asserted as expected, a success message is displayed and the "for" loop code starts executing. If the signal "memory_ready" is not asserted as expected, then a failure message is displayed and the simulation exits.

1.37 SVA using local variables

A variable can be declared locally within a sequence or a property and an assignment can be made on that variable. The variable is placed next to a sub-sequence separated by a comma. If the sub-sequence matches, then the variable assignment is executed. Every time the sequence is attempted, a new copy of the variable is created.

```
property p_local_var1;
int lvar1;
@(posedge clk)
($rose(enable1), lvar1 = a) |->
          ##4 (aa == (lvar1*lvar1*lvar1));
endproperty

a_local_var1: assert property(p_local_var1);
```

The property p_local_var1 looks for a rising edge on the signal "enable1." Upon a match on this, the local variable "lvar1" stores the value of the design vector "a." After 4 cycles, it is checked that the value of the design output vector "aa" is equal to the cubed value of the local variable. The consequent of the property waits for the design to satisfy the latency (4 clock cycles) and then compares the original design output with the locally calculated value. Figure 1-41 shows how the check reacts in a simulation.

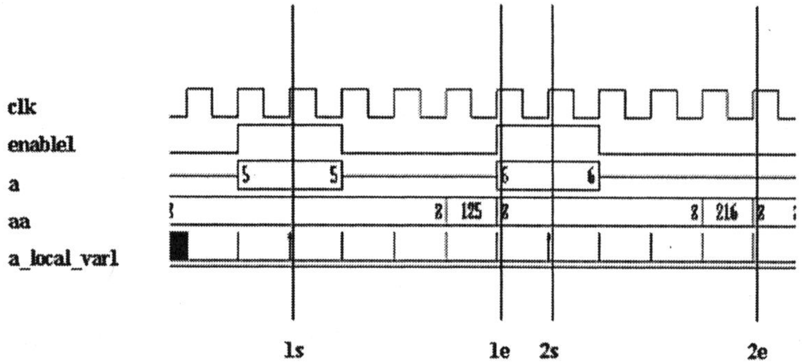

Figure 1-41. Waveform for SVA with local variables

The marker 1s shows the point where the rising edge of the signal "enable1" is sampled. At this point, vector "a" has a value of 5 and this is stored in the local variable "lvar1." The marker 1e shows the point where the output is sampled. This is 4 clock cycles after the input value was stored. At marker 1e, since the output value (125) equals that of the cube of the local variable "lvar1," the assertion succeeds. Similarly, marker 2s shows when the next input data is stored and marker 2e shows when the output is sampled and compared with the cubed value of local variable "lvar1."

The local variables can be stored and manipulated inside SVA.

```
property p_lvar_accum;
int lvar;
@(posedge clk) $rose(start)|=>
(enable1 ##2 enable2, lvar = lvar +aa) [*4]
##1 (stop && (aout == lvar));
endproperty

a_lvar_accum : assert property(p_lvar_accum);
```

The property p_lvar_accum checks for the following.

1. A valid start occurs if a rising edge is detected on the signal "start" on any given positive edge of the clock.
2. One cycle later, a specific pattern or a sub-sequence is looked for. The signal "enable1" should be detected high and 2 cycles later signal "enable2" should be detected high. This sub-sequence should repeat itself 4 times continuously.
3. For every repeat of the sub-sequence, the value of the vector "aa" is accumulated locally. At the end of the repetition, the local variable holds a value accumulated from the vector "aa" four times.
4. One cycle after the repetition, it is expected that the signal "stop" is detected high and the value held by the local variable is equal to the value held by the output vector "aout."

Figure 1-42 shows how the check reacts in a simulation.

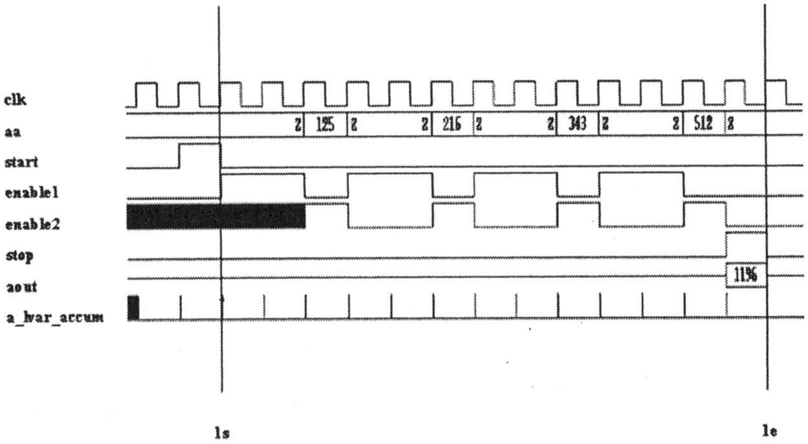

Figure 1-42. SVA with local variable assignment

The marker 1s shows a valid start of the check wherein the signal "start" is detected high. Marker 1e shows the end of the check. The repetitions of the enable signals complete successfully and one cycle later the signal "stop" is detected high as expected. The local variable holds the same value as that of the output vector "aout" and hence the check succeeds at the marker 1e.

1.38 SVA calling subroutine on a sequence match

SVA can also call a subroutine on every successful match of a sequence. The local variables defined in the same sequence can be passed as arguments to these subroutine calls. For each match on the sequence, the subroutine calls are executed in the same order as they are listed in the sequence definition.

```
sequence s_display1;
@(posedge clk)
($rose(a), $display("Signal  a  arrived  at  %t\n",
$time));
endsequence

sequence s_display2;
@(posedge clk)
($rose(b), $display("Signal  b  arrived  at  %t\n",
$time));
endsequence

property p_display_window;
@(posedge clk)
s_display1 |-> ##[2:5] s_display2;
endproperty

a_display_window :
      assert property(p_display_window);
```

Sequence s_display1 looks for a rising edge on the signal "a." Upon a match on this event, it executes the display statement. Sequence s_display2 does a similar action on signal "b." The property p_display_window checks that if sequence s_display1 occurs then the sequence s_display2 should occur anywhere between 2 and 5 clock cycles. By using display statements, the user can get information on exactly how many cycles the consequent sequence completed. Figure 1-43 shows how the check reacts in a simulation.

The marker 1s shows a valid start of the checker since a rising edge of the signal "a" is detected. At this point, SVA executes the display statement relevant to this sequence (s_display1). The marker 1e shows the point when a rising edge arrives on signal "b." Since this arrives after 3 cycles, the checker succeeds. At this point, the display statement relevant to this sequence (s_display2) is executed.

The marker 2s shows a valid start of the checker since a rising edge of the signal "a" is detected. At this point, SVA executes the display statement relevant to this sequence (s_display1). The marker 2e shows the ending point of the checker. A valid rising edge never arrived on signal "b" within 2 and 5 clock cycles and hence, the checker failed. Since the second sequence never matched, the relevant display statement is not executed. A default error is issued by SVA.

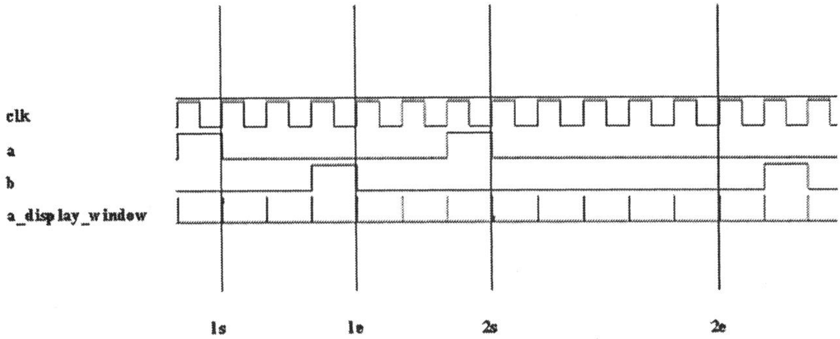

Figure 1-43. SVA using subroutines on sequence match

A sample simulation log is shown below.

```
Signal a arrived at                      125

Signal b arrived at                      275

"sub.v", 45: sub.a_display_window:
started at 125s succeeded at 275s

Signal a arrived at                      425

"sub.v", 45: sub.a_display_window:
started at 425s failed at 675s
        Offending '$rose(b)'
```

1.39 Connecting SVA to the design

SVA checkers can be connected to the design by two different methods.

1. Embed or in-line the checkers in the module definition.
2. Bind the checkers to a module, an instance of a module or multiple instances of a module.

Some engineers don't like adding any verification code within the design. In this case, binding the SVA checkers externally is the choice. SVA code can be embedded anywhere in a module definition. The following example shows SVA being in-lined within the module.

```
module inline(clk, a, b, d1, d2, d);

input logic clk, a, b;
input logic [7:0] d1, d2;
output logic [7:0] d;

always@(posedge clk)
begin
  if(a)
    d <= d1;
  if(b)
    d <= d2;
end

property p_mutex;
  @(posedge clk) not (a && b);
endproperty

a_mutex: assert property(p_mutex);

endmodule
```

If the user decides to keep the SVA checkers separate from the design, then he has to create a separate checker module. ***By defining a separate checker module, the re-usability of the checker increases. The following code shows a checker module.***

```
module mutex_chk(a, b, clk);
```

```
input logic a, b, clk;

property p_mutex;
 @(posedge clk) not (a && b);
endproperty

a_mutex: assert property(p_mutex);

endmodule
```

Note that when a checker module is defined, it is an independent entity. The checker is written for a generic set of signals. The checker can be bound to any module or instance in the design. The syntax for binding is as follows.

```
bind <module_name or instance name>
     <checker name> <checker instance name>
     <design signals>;
```

For the example checker shown above, the binding can be done as follows.

```
bind inline mutex_chk i2 (a, b, clk);
```

When the binding is done, the actual design signal names are used.

Let's say we have a top-level module as follows.

```
module top (..);

inline u1 (clk, a, b, in1, in2, out1);
inline u2 (clk, c, d, in3, in4, out2);

endmodule
```

The checker mutex_chk can be bound to the two instances of the module "inline" in the top-level module as follows.

```
bind top.u1 mutex_chk i1(a, b, clk);
bind top.u2 mutex_chk i2(c, d, clk);
```

The design signals that are bound can contain cross module reference to any signal within the scope of the bound instance.

1.40 SVA for functional coverage

Functional coverage is a metric for measuring verification status against design specification. It is classified into two categories:

a. Protocol coverage.
b. Test plan coverage.

Assertions can be used to get exhaustive information on protocol coverage. SVA provides a keyword "**cover**" to specify this. The basic syntax of a **cover** statement is as follows.

```
<cover_name> : cover property(property_name)
```

"cover_name" is a name provided by the user to identify the coverage statement and "property_name" is the name of the property on which the user wants to get coverage information. For example, the checker "mutex_chk" defined in Section 1.39 can be covered as follows.

```
c_mutex: cover property(p_mutex);
```

The results of the **cover** statement will provide the following information:

1. Number of times the property was attempted.
2. Number of times the property succeeded.
3. Number of times the property failed.
4. Number of times the property succeeded vacuously.

A sample coverage log from a simulation for the checker "mutex_chk" is shown below.

```
c_mutex, 12 attempts, 12 match, 0 vacuous match
```

Just like the assert statement, the cover statement can also have an action block. Upon a successful coverage match, a function or a task can be called or a local variable update can be performed.

Chapter 2

SVA SIMULATION METHODOLOGY

In Chapter 1, SVA language constructs were discussed in detail with examples. All examples were illustrated as relationships between two or more generic signals without any design details. In Chapter 2, a dummy system is used to present a real situation. The process of protocol extraction and assertion development will be discussed step by step. Various simulation methodologies that can significantly increase the productivity of assertion based verification will be discussed. Functional coverage and reactive testbench development will be discussed in detail.

2.1 A sample system under verification

The sample system under consideration is shown in Figure 2-1. The system has 3 master devices and 2 target devices. A link is established between the master and the target devices by the mediator. At a given time, only one master can conduct a transaction and with only one target device. Any master device can conduct a transaction with any target device. The transaction can be a read or a write. The mediator contains arbiter logic that decides which master will be allowed to conduct a transaction. The arbiter uses a simple round robin technique. The mediator also contains glue logic that actually decodes the master information for the target device and vice versa. The glue logic helps establish the link between a specific master device and target device to conduct the transaction successfully.

2.1.1 The Master device

The block diagram of the master device along with input and output ports is shown in Figure 2-2. The master device can perform a read and a write transaction. It can support 2 target devices in a single system. When the master device gets the instruction "ask_for_it," it is ready to perform a transaction. It sends an active low pulse on the "req" signal and waits for a "gnt." The "gnt" signal is an active low signal. If the "gnt" signal does not come within 2 to 5 clock cycles, then the master will retry to get access at a later time. If the "gnt" is acquired, then the master will immediately assert the "frame" and "irdy" signals acknowledging the arrival of the "gnt" signal ("frame" and "irdy" are active low signals). In the same clock cycle it also selects the target device it will have the transaction with. The master uses the output signal "rsel" to indicate this. If signal "rsel" is set to 1, then the master will to have a transaction with target device 1. If the signal "rsel" is set to 0, then the master will have a transaction with target device 0.

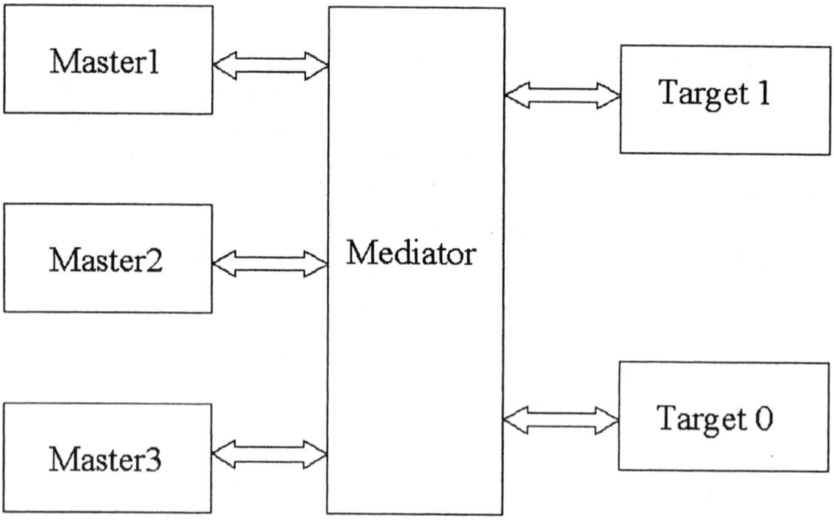

Figure 2-1. A sample system

Once the signal "rsel" is updated, the target device is expected to identify itself to the master. The target device uses the signal "trdy" to acknowledge its readiness. If the target does not acknowledge itself within 3 clock cycles from the point when "rsel" is assigned, it is an error condition. If the target does acknowledge itself, then the master decides whether to read or write. The master sends the data and the instruction whether to read or write through the "datac" bus.

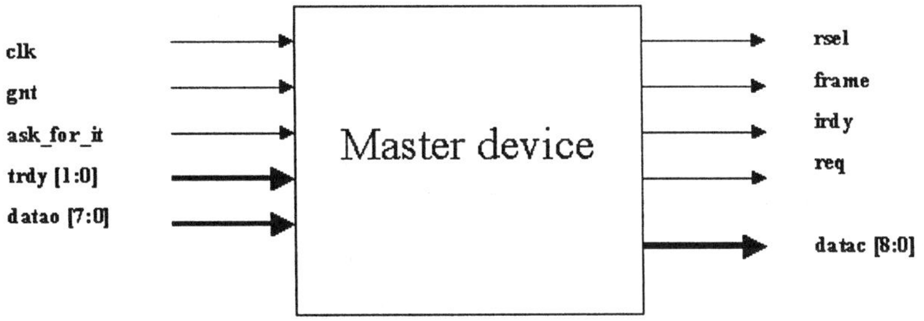

Figure 2-2. Sample master device

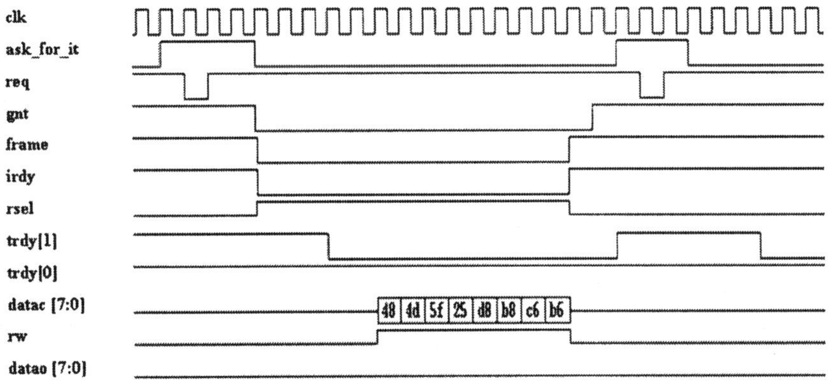

Figure 2-3. Write transaction of a master device

The most significant bit is the instruction bit (shown as signal "rw" in waveforms). If it is 1, the master will write and if it is a 0 then the master will read. If it is a write transaction, the least significant 8 bits consist of the data that needs to be written to the target device. If it is a read transaction, then the data read from the target device appears on the "datao" input bus. Each transaction of the master will last exactly 8 clock cycles. In other words, a master can either read 8 bytes in a transaction or write 8 bytes in a transaction. There is no specific address generation scheme. The master will write to the most updated write pointer address existing within the target device. Similarly, the master will read from the most updated read pointer address within the target device. The sample waveform for a master write transaction is shown in Figure 2-3. The sample waveform for a master read transaction is shown in Figure 2-4.

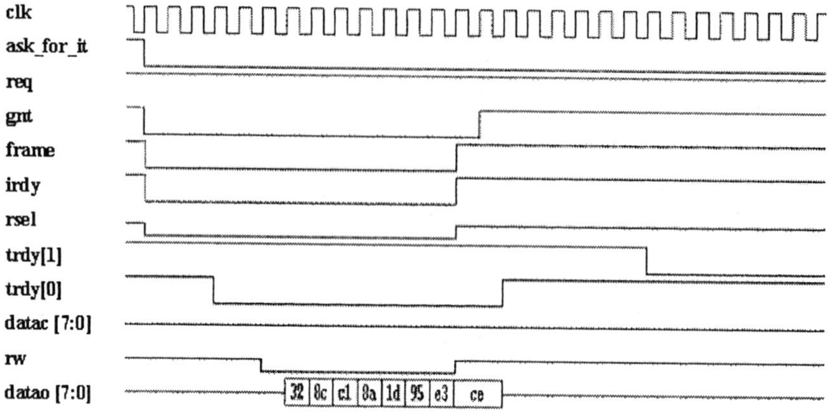

Figure 2-4. Sample read transaction of a master device

Once the read or write transaction is complete, the master indicates completion by de-asserting the signals "frame" and "irdy" in the next clock cycle. It also sets the "rsel" signal to tri-state. The arbiter acknowledges this and de-asserts the "gnt" signal in the next clock cycle. Once the arbiter removes the "gnt" signal, the target device acknowledges completion of the transaction by de-asserting the "trdy" signal.

2.1.2 The Mediator

The block diagram of the mediator along with input and output ports is shown in Figure 2-5. The mediator performs two important tasks:

1. Provide arbitration logic that decides which master will get access to conduct a transaction with a target device.

2. Establish the link between a specific master device and a target device. At a given time any number of masters can ask for access by asserting their respective "req" signal.

The arbiter uses a round robin algorithm and decides which master will get access. When the arbiter makes a decision, it will assert the "gnt" signal of the respective master device. The arbiter can take anywhere between 2 to 5 clock cycles to make a decision. The internal logic for the arbiter is described with a simple zero one-hot state machine.

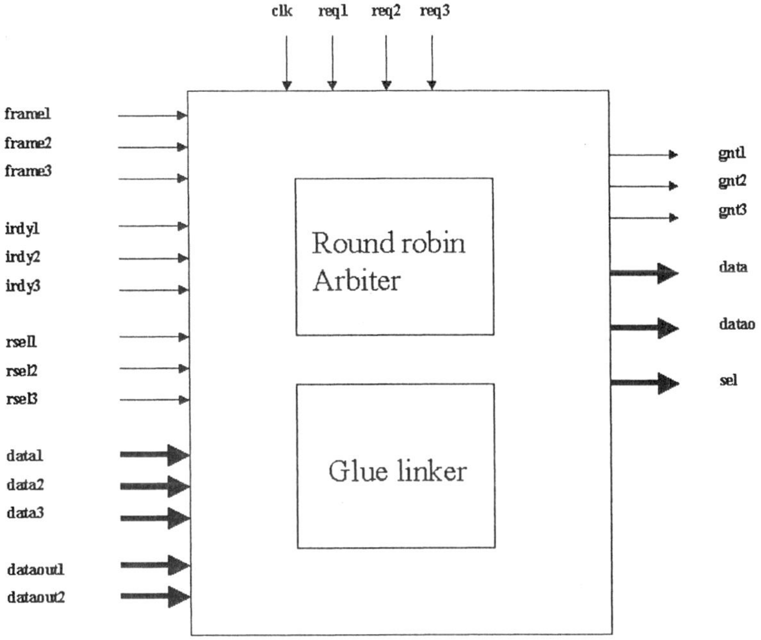

Figure 2-5. Sample mediator device

After the master selects the target it will have a transaction with, the mediator will provide that information to the specific target device. Since three masters are capable of having a transaction with any of the target devices, the mediator has to monitor the "rsel" signals from all three masters. At any given time, either all the three "rsel" signals are tri-stated or definitely two of them are tri-stated. If all three "rsel" signals are tri-stated, then there is no transaction request at that point. If there is a transaction, then one of the "rsel" signals will have a value of 0 or 1, depending on which target device will be used. If signal "rsel" is 1 then, the MSB of signal "sel" is set high indicating that target device 1 is selected. If signal "rsel" is 0 then, the LSB of signal "sel" is set high indicating that target device 0 is selected.

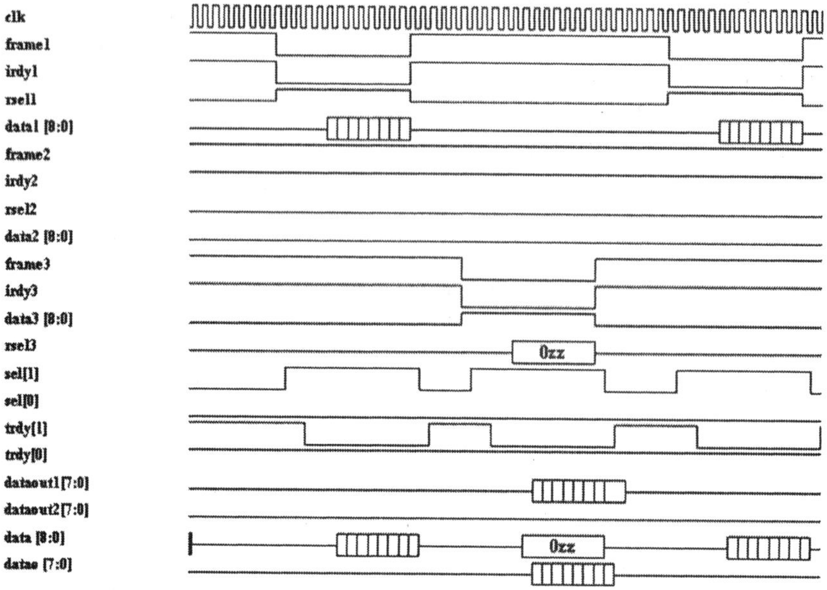

Figure 2-6. Waveform for mediator functionality

The mediator also selects the correct data signals for both write and read transactions. If it is a write transaction, then the mediator monitors which master's "rsel" signal is active and assigns the data value relevant to that master to the selected target device input. For example, if master 1 is asking for a write transaction with target device zero, then the signal "rsel1" will be set to low and the bus "data1" will be assigned to the mediator output bus "data." This output is fed to the input of the selected target device. The mediator also assigns the correct output data from the target device back to the master device in a read transaction. For example, if target 1 is involved in the read transaction, then the bus "dataout1" will be assigned to the bus "datao." The sample waveform for the mediator is shown in Figure 2-6.

2.1.3 The Target device

The block diagram of the target device along with input and output ports is shown in Figure 2-7. The target device has a first-in-first-out type memory that can store up to 64 bytes of data.

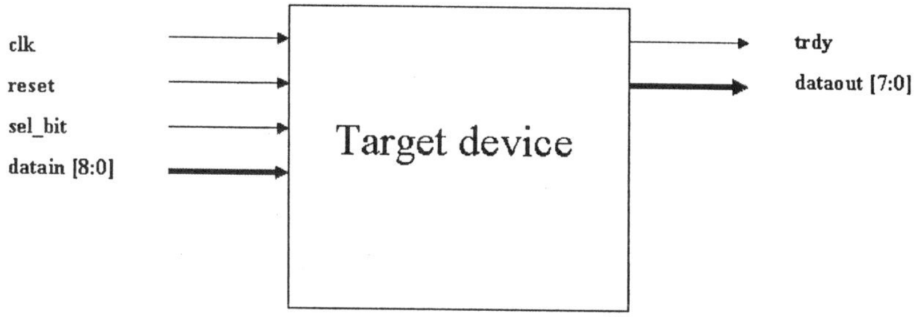

Figure 2-7. Sample target device

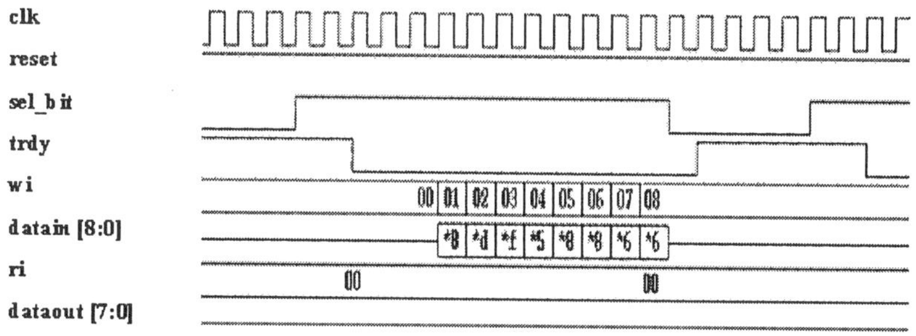

Figure 2-8. Target write transaction

The target device waits for the signal "sel_bit" to be asserted. Once signal "sel_bit" is asserted, the target has to acknowledge by asserting the signal "trdy" after 2 clock cycles. After asserting signal "trdy" the target device waits for a valid data and a valid write signal if it is a write transaction. Once a valid write signal is detected, the incoming data is stored in the target device in locations starting from the most updated value of the write pointer (wi) register. If it is a read transaction, then the target device reads out 8 data points from its memory using the current read pointer location (ri) as the starting address.

The type of transaction is indicated by the MSB of the bus "datain." In a read transaction, the data read appears on the bus vector "dataout." When the transaction is complete, the signal "sel_bit" is de-asserted and one clock cycle after that the signal "trdy" is de-asserted. The sample waveform for a target write transaction is shown in Figure 2-8. The sample waveform for a target read transaction is shown in Figure 2-9.

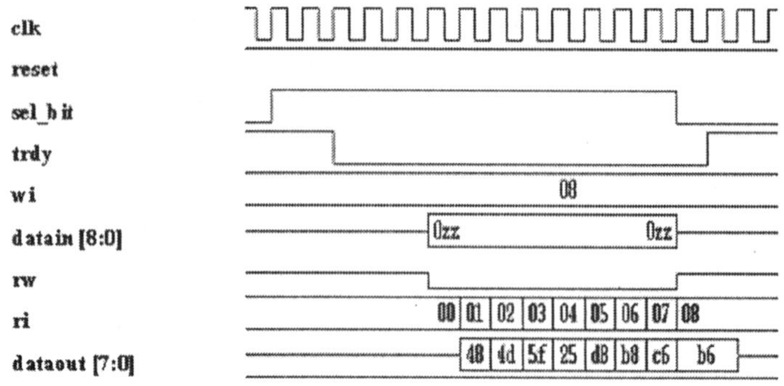

Figure 2-9. Target read transaction

2.2 Block level verification

As the individual design blocks get ready they should be tested thoroughly. Exhaustive verification of the blocks will uncover the corner case bugs ahead of time. Finding these bugs before integrating the system is a must. Finding these bugs at the system level will be very difficult. Also, system level failures provide a greater challenge for identifying and debugging corner case bugs. SVA can be used efficiently to test the individual blocks effectively. ***At the block level, the simulations are smaller and hence the bugs can be traced easily and fixed promptly***. There are 4 individual design blocks in the sample system that need to be verified:

1. Master
2. Target
3. Arbiter
4. Glue

There are also 2 block level interfaces that need to be tested thoroughly:

1. Master and Mediator
2. Target and Mediator

2.2.1 SVA in design blocks

The following tips are recommended for doing block level verification with SVA:

- All SVA checks written for a block level design should be in-lined. Block level assertions often involve accessing internal registers of a design and hence, in-lining the checks within the design module is more efficient.

- The inclusion of SVA checks written at the block level should be controlled by a parameter defined within the design module. This gives the freedom to turn the checks on and off on a per simulation basis.

- The severity level of the SVA checks written at the block level should be controlled by a parameter defined within the design module. The default severity in SVA is to print an error message and continue simulating.

- Every block level SVA check written should be asserted and covered. It is a must that all the block level checks must have at least one real success.

2.2.2 Arbiter verification

Based on the protocol description of the arbiter from Section 2.1.2, the following SVA checks can be extracted. Some of the common expressions used repeatedly in the arbiter checks can be defined with "assign" statements as shown below:

```
assign frame = frame1 && frame2 && frame3;
assign irdy = irdy1 && irdy2 && irdy3;
assign gnt = !gnt1 || !gnt2 || !gnt3;
assign req = !req1 || !req2 || !req3;
```

The "frame" and "irdy" signals are all active low signals. Each master has a unique "frame" and "irdy" signal and these are inputs to the arbiter module. If a master is active, it sets both the "frame" and "irdy" low. Hence, by AND'ing the "frame" signals, we know that the bus is active if the AND'ed value is low. Similarly, by AND'ing the "irdy" signals, we know that the bus is active if the AND'ed value is low. If the AND'ed values of "frame" and "irdy" signals are high, then none of the masters are active.

Each master has a unique "req" signal that requests the bus and the arbiter provides a unique "gnt" signal. By OR'ing all the "req" signals we know that even if one master has a valid request, the arbiter considers the request. Similarly, by OR'ing the "gnt" signals, we know that one master has acquired the grant. ***Creating such intermediate expressions make the SVA checkers more readable.***

Arb_chk1: On any given clock edge, the internal state of the arbiter should behave as a zero one-hot state machine.

```
property p_arb_onehot0;
  @(posedge clk) $onehot0(state);
endproperty
```

Arb_chk2: Upon a valid request by a master, the arbiter should provide a grant within 2 to 5 clock cycles.

```
property p_req_gnt;
  @(posedge clk) $rose (req) |->
                ##[2:5] $rose (gnt);
endproperty
```

Arb_chk3: Once the grant is awarded, the master should acknowledge acceptance in the same clock cycle by asserting the "frame" and "irdy" signals.

```
property p_gnt_frame;
  @(posedge clk) $rose (gnt) |->
                     $fell (frame && irdy);
endproperty
```

Arb_chk4: Once the master completes the transaction it de-asserts the "frame" and "irdy" signals, followed by that, the arbiter should de-assert the "gnt" signal on the next clock cycle.

```
property p_frame_gnt;
  @(posedge clk) $rose(frame && irdy)
                           |=> $fell(gnt);
Endproperty
```

2.2.3 SVA Checks for arbiter in simulation

The four checks shown in Section 2.2.2 should be in-lined within the arbiter module. There should be a provision to assert these properties on a need basis. The following code shows how this can be achieved.

```
module arbiter(.....);

// port declarations

parameter arb_sva = 1'b1;
parameter arb_sva_severity = 1'b1;

// Arbiter design description
// SVA property description

// SVA Checks

always@(posedge clk)
begin
if(arb_sva)
begin

a_arb_onehot0:
     assert property(p_arb_onehot0)
     else if(arb_sva_severity) $fatal;

a_req_gnt:
     assert property(p_req_gnt)
     else if(arb_sva_severity) $fatal;

a_gnt_frame :
     assert property(p_gnt_frame)
     else if(arb_sva_severity) $fatal;

a_frame_gnt:
     assert property(p_frame_gnt)
     else if(arb_sva_severity) $fatal;

c_arb_onehot0: cover property(p_arb_onehot0);
c_req_gnt: cover property(p_req_gnt);
c_gnt_frame: cover property(p_gnt_frame);
c_frame_gnt: cover property(p_frame_gnt);

end
end

endmodule
```

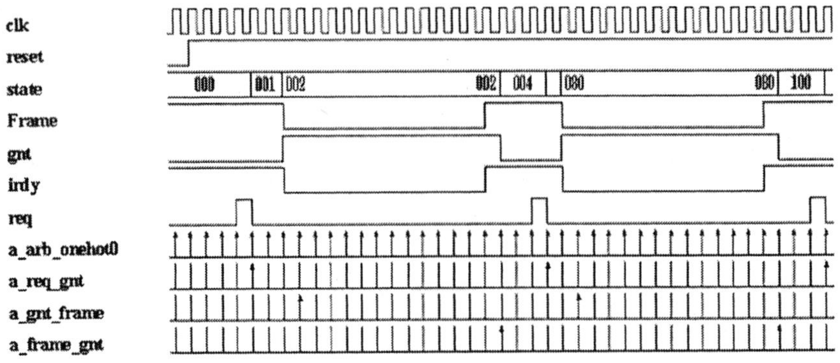

Figure 2-10. Arbiter checks in simulation

The parameter "arb_sva" will have to be set to 1 for the checks to be included in a simulation. The parameter "arb_sva_severity" controls the action to be taken during simulation. In this case, if the parameter is set to 1, then the severity is set to $fatal. This means that upon a failure of any of these checks, the simulation will exit. By setting the parameter to 0, the checks will use the default condition, which is to print an error message on a failure and continue simulating. A waveform from a sample simulation is shown in Figure 2-10.

2.2.4 Master verification

Based on the protocol description of the master from Section 2.1.1, the following SVA checks can be extracted. Note that each master has only one "req," "gnt," "frame" and "irdy" signals. The mention of these signals in the master checkers does not represent the expressions defined in the arbiter checkers. They are just individual signals present in each master device.

Master_chk1: Upon a valid request from a master, the grant shall come within 2 to 5 clock cycles. If so and if the signal "r_sel" is high, then on the same clock cycle, the master should assert the signals "frame" and "irdy." Three cycles later the target device one should acknowledge its selection by asserting the signal "trdy."

```
property p_master_start1;
@(posedge clk)
```

```
($fell (req) ##[2:5] ($fell(gnt) && r_sel)) |->
              (!frame && !irdy) ##3 !trdy[1];
endproperty
```

Master_chk2: Upon a valid request from a master, the grant shall come within 2 to 5 clock cycles. If so and if the signal "r_sel" is low, then on the same clock cycle, the master should assert the signals "frame" and "irdy." Three cycles later the target device zero should acknowledge its selection by asserting the signal "trdy."

```
property p_master_start2;
  @(posedge clk)
  ($fell (req) ##[2:5] ($fell(gnt) && !r_sel))|->
              (!frame && !irdy) ##3 !trdy[0];
endproperty
```

Master_chk3: Once the target acknowledges its selection, the master should complete its transaction within 10 clock cycles. It should indicate the transaction completion by de-asserting the signals "frame" and "irdy." One cycle later the signal "gnt" should be de-asserted.

```
property p_master_stop1;
  @(posedge clk)
$fell (trdy[1]) |-> ##10 (frame && irdy) ##1 gnt;
endproperty

property p_master_stop2;
  @(posedge clk)
$fell (trdy[0]) |-> ##10 (frame && irdy) ##1 gnt;
endproperty
```

Note that two separate properties are written to check the transaction completion, one for each target device.

Master_chk4: If the master is in a write transaction, then the bus data (data_c) should not be tri-stated and should have valid data.

```
property p_master_data1;
  @(posedge clk)
    ($fell (trdy[1]) ##2 rw) |->
        ($isunknown(data) == 0) [*7];
endproperty
```

```
property p_master_data2;
  @(posedge clk)
     ($fell (trdy[0]) ##2 rw) |->
        ($isunknown(data) == 0) [*7];
endproperty
```

- Note that two separate properties are written to check the validity of data during write transaction, one for each target device.
- Note that if the signal "rw" is high, then the master is conducting a write transaction.

Master_chk5: If the master is in a read transaction, then the bus data (data_o) should not be tri-stated and should have valid data.

```
property p_master_datao1;
  @(posedge clk)
       ($fell (trdy[1]) ##3 !rw) |=>
            ($isunknown(data_o) == 0) [*7];
endproperty

property p_master_datao2;
  @(posedge clk)
       ($fell (trdy[0]) ##3 !rw) |=>
          ($isunknown(data_o) == 0) [*7];
endproperty
```

- Note that two separate properties are written to check the validity of data during read transaction, one for each target device.
- Note that if the signal "rw" is low, then the master is conducting a read transaction.

2.2.5 SVA Checks for the master in simulation

The five checks shown in Section 2.2.4 should be in-lined within the master module. There should be a provision to assert these properties on a need basis. The following code shows how this can be achieved.

```
module master(.....);

// port declarations

parameter master_sva = 1'b1;
parameter master_sva_severity = 1'b1;
```

```
// Master design description

// SVA property description

// SVA Checks

always@(posedge clk)

begin

if(master_sva)

begin

a_master_start1:
    assert property(p_master_start1)
    else if(master_sva_severity) $fatal;

a_master_start2:
    assert property(p_master_start2)
    else if(master_sva_severity) $fatal;

a_master_stop1:
    assert property(p_master_stop1)
    else if(master_sva_severity) $fatal;

a_master_stop2:
    assert property(p_master_stop2)
    else if(master_sva_severity) $fatal;

a_master_data1:
    assert property(p_master_data1)
    else if(master_sva_severity) $fatal;

a_master_data2:
    assert property(p_master_data2)
    else if(master_sva_severity) $fatal;

a_master_datao1:
    assert property(p_master_datao1)
    else if(master_sva_severity) $fatal;
```

```
a_master_datao2:
    assert property(p_master_datao2)
    else if(master_sva_severity) $fatal;

c_master_start1: cover property(p_master_start1);
c_master_start2: cover property(p_master_start2);
c_master_stop1: cover property(p_master_stop1);
c_master_stop2: cover property(p_master_stop2);
c_master_data1: cover property(p_master_data1);
c_master_data2: cover property(p_master_data2);
c_master_datao1: cover property(p_master_datao1);
c_master_datao2: cover property(p_master_datao2);

end

end

endmodule
```

A waveform from a sample simulation of these master checks is shown in Figure 2-11.

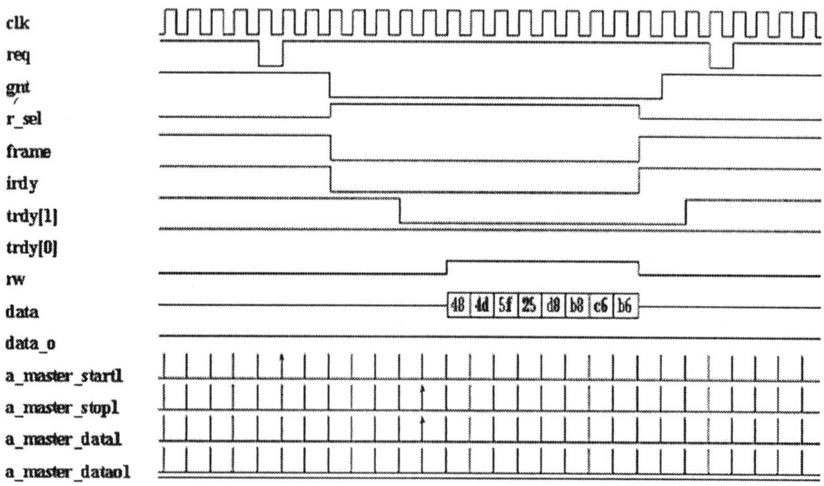

Figure 2-11. Master checks in simulation for target 1

2.2.6 Glue verification

Based on the protocol description of the glue logic from Section 2.1.2, the following SVA checks can be extracted.

Glue_chk1: If any one of the master select signals "sel1," "sel2" or "sel3" is high, then target device one should be selected.

```
property p_sel_1;
  @(posedge clk)
    (rsel1 || rsel2 || rsel3) |=> sel == 2'b10;
endproperty
```

Glue_chk2: If any one of the master select signals "sel1," "sel2" or "sel3" is low, then target device zero should be selected.

```
property p_sel_0;
  @(posedge clk)
    (!rsel1 || !rsel2 || !rsel3) |=> sel == 2'b01;
endproperty
```

Glue_chk3: During a write transaction, if the signal "rsel1" is not tri-stated, then the data from master device one should be written to the respective target device.

```
property p_rsel1_write;
  @(posedge clk)
((rsel1 || !rsel1) ##3 ($fell (trdy[1]) ||
$fell(trdy[0])) ##3 data1[8]) |->
      (data == $past(data1)) [*7];
endproperty
```

- Note that we determine the nature of the transaction (read/write) by using the most significant bit of the bus "data."
- If the MSB of the bus "data" is high, then it is a write transaction.
- If the MSB of the bus "data" is low, then it is a read transaction.
- Within the master device, the nature of the transaction is determined by the signal "rw." This signal is a copy of the MSB of the bus "data." The signal "rw" is local to the master device. The external interface should infer the nature of the transaction by using the MSB of the bus "data."

Glue_chk4: During a write transaction, if the signal "rsel2" is not tri-stated, then the data from master device two should be written to the respective target device.

```
property p_rsel2_write;
 @(posedge clk)
((rsel2   ||   !rsel2)   ##3   ($fell   (trdy[1])   ||
$fell(trdy[0])) ##3 data2[8]) |->
       (data == $past(data2)) [*7];
endproperty
```

Glue_chk5: During a write transaction, if the signal "rsel3" is not tri-stated, then the data from master device three should be written to the respective target device.

```
property p_rsel3_write;
 @(posedge clk)
((rsel3   ||   !rsel3)   ##3   ($fell   (trdy[1])   ||
$fell(trdy[0])) ##3 data3[8]) |->
         (data == $past(data3)) [*7];
Endproperty
```

Glue_chk6: During a read transaction, if target device one is selected, then data read from target one (dataout1) should be fed back to the respective master.

```
property p_read1;
 @(posedge clk)
($fell (trdy[1]) ##4 !data[8]) |->
               (dataout1 == datao) [*7];
endproperty
```

Glue_chk7: During a read transaction, if target device zero is selected, then data read from target zero (dataout2) should be fed back to the respective master.

```
property p_read0;
 @(posedge clk)
($fell (trdy[0]) ##4 !data[8]) |->
               (dataout2 == datao) [*7];
endproperty
```

2.2.7 SVA Checks for the glue logic in simulation

The seven checks shown in Section 2.2.6 should be in-lined within the glue module. There should be a provision to assert these properties on a need basis. The following code shows how this can be achieved.

```
module glue(.....);

// port declarations

parameter glue_sva = 1'b1;
parameter glue_sva_severity = 1'b1;

// glue design description
// glue SVA property description

// SVA Checks

always@(posedge clk)
begin
if(glue_sva)
begin

a_sel_1:
     assert property(p_sel_1)
     else if(glue_sva_severity) $fatal;

a_sel_0:
     assert property(p_sel_0)
     else if(glue_sva_severity) $fatal;

a_rsel1_write:
     assert property(p_rsel1_write)
     else if(glue_sva_severity) $fatal;

a_rsel2_write:
     assert property(p_rsel2_write)
     else if(glue_sva_severity) $fatal;

a_rsel3_write:
     assert property(p_rsel3_write)
     else if(glue_sva_severity) $fatal;
```

```
a_read1:
     assert property(p_read1)
     else if(glue_sva_severity) $fatal;

a_read0:
     assert property(p_read0)
     else if(glue_sva_severity) $fatal;

c_sel_1: cover property(p_sel_1);
c_sel_0: cover property(p_sel_0);
c_rsel1_write: cover property(p_rsel1_write);
c_rsel2_write: cover property(p_rsel2_write);
c_rsel3_write: cover property(p_rsel3_write);
c_read1: cover property(p_read1);
c_read0: cover property(p_read0);

end
end

endmodule
```

A waveform from a sample simulation of the glue checks is shown in Figure 2-12.

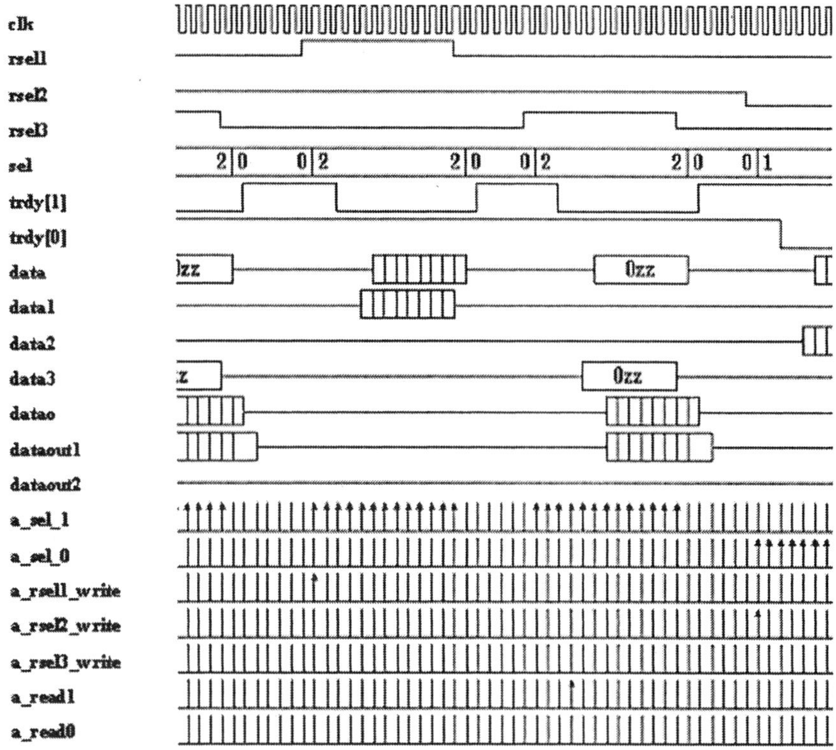

Figure 2-12. Glue checks in simulation

2.2.8 Target verification

Based on the protocol description of the target device from Section 2.1.3, the following SVA checks can be extracted.

Target_chk1: If a target is selected, then it should assert the signal "trdy" after 2 clock cycles.

```
property p_sel_trdy_start;
 @(posedge clk) $rose (sel_bit) |->
                   ##1 trdy ##1 !trdy;
endproperty
```

Target_chk2: At the end of a transaction, the "sel_bit" signal is de-asserted. One clock cycle after that, the signal "trdy" should be de-asserted.

```
property p_sel_trdy_stop;
  @(posedge clk) $fell (sel_bit) |=> trdy;
endproperty
```

Target_chk3: In a write transaction, the write pointers should be incremented by one after each clock cycle to complete a valid "write" to a unique address every time.

```
property p_write;
  @(posedge clk)
 (datain[8] && sel_bit && (wi != 0)) |->
                     (wi == ($past(wi) + 1));
endproperty
```

- Note that the address pointer will roll over from 63 to 0. Hence, this check cannot be applied if on a given clock edge the write pointer is at 0.
- A different check can be written to verify that the pointer always rolls over correctly from 63 to 0.

Target_chk4: In a read transaction, the read pointers should be incremented by one after each clock cycle to complete a valid "read" from a unique address every time.

```
property p_read;
  @(posedge clk)
 (!datain[8] && sel_bit && (ri != 63)) |=>
                     (ri == ($past(ri) + 1));
endproperty
```

- Note that in the case of read pointer, when the pointer is at 63 this check cannot be applied.
- The read operation has a latency of one clock cycle and hence we use the Non-overlapping implication operator.
- Since a non-overlapping operator is used, the check moves forward to one cycle and compares the address in the previous cycle.
- For example, on a given clock edge, if the antecedent of the implication is true, the check moves to the next clock cycle. If the pointer is at 63, then the check moves to pointer 0 and compares 63 and 0 for an increment of one. This is incorrect. Hence, the check should not be performed if the value of the read pointer is 63 on a given clock edge.

- A separate check can be written to make sure that the pointer rolls over from 63 to 0 accurately.

Target_chk5: During a valid read or write transaction, the data read from or written to the target should be valid.

```
property p_target_datain;
  @(posedge clk)
   ($fell (trdy) ##3 (datain[8])) |->
            not ($isunknown (datain)) [*7];
endproperty

property p_target_dataout;
  @(posedge clk)
  ($fell (trdy) ##3 (!datain[8])) |=>
          not ($isunknown(dataout)) [*7];
endproperty
```

2.2.9 SVA Checks for the Target in simulation

The five checks shown in Section 2.2.8 should be in-lined within the target module. There should be a provision to assert these properties on a need basis. The following code shows how this can be achieved.

```
module target(.....);

// port declarations

parameter target_sva = 1'b1;
parameter target_sva_severity = 1'b1;

// target design description
// target SVA property description
// SVA Checks

always@(posedge clk)
begin
if(target_sva)
begin

a_sel_trdy_start:
    assert property(p_sel_trdy_start)
```

```
    else if(target_sva_severity) $fatal;
a_sel_trdy_stop:
    assert property(p_sel_trdy_stop)
    else if(target_sva_severity) $fatal;

a_write:
      assert property(p_write)
      else if(target_sva_severity) $fatal;

a_read:
      assert property(p_read)
      else if(target_sva_severity) $fatal;

a_target_datain:
      assert property(p_target_datain)
      else if(target_sva_severity) $fatal;

a_target_dataout:
      assert property(p_target_dataout)
      else if(target_sva_severity) $fatal;

c_sel_trdy_start:
                cover property(p_sel_trdy_start);
c_sel_trdy_stop: cover property(p_sel_trdy_stop);
c_write: cover property(p_write);
c_read: cover property(p_read);
c_target_datain: cover property(p_target_datain);
c_target_dataout:
                cover property(p_target_dataout);

end
end
endmodule
```

A waveform from a sample simulation of the target checks is shown in Figure 2-13.

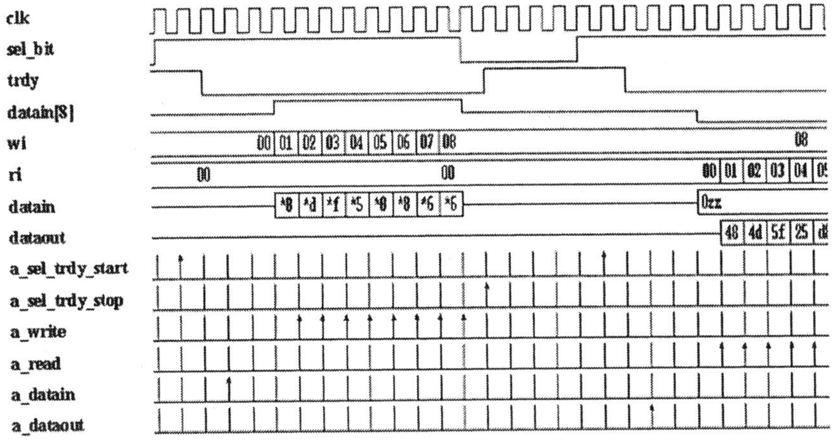

Figure 2-13. Target checks in simulation

2.3 System level verification

There are 3 masters and 2 targets in the system along with an instance of the mediator. The top-level connection of the system is shown below.

```
Module top(..,..,    );

// port declarations

master u1 (ask[2], clk, req1, gnt1, frame1,
irdy1, trdy, data1, rsel1, datao);

master u2 (ask[1], clk, req2, gnt2, frame2,
irdy2, trdy, data2, rsel2, datao);

master u3 (ask[0], clk, req3, gnt3, frame3,
irdy3, trdy, data3, rsel3, datao);

arbiter u4 (clk, reset, frame, irdy, req1, req2,
req3, gnt1, gnt2, gnt3);

glue u5 (clk, frame1, irdy1, frame2, irdy2,
frame3, irdy3, trdy, rsel1, rsel2, rsel3, data1,
```

```
data2, data3, sel, data, dataout1, dataout2,
datao);

target u6 (clk, reset, sel[1], trdy[1], data,
dataout1);

target u7 (clk, reset, sel[0], trdy[0], data,
dataout2);

endmodule
```

The following tips are recommended for doing system level verification with SVA:

- Since the internal functionality of the individual blocks was verified thoroughly, the block level assertions don't have to be included during the system level verification by default. The main motive behind this is performance.

- If performance is not a bottleneck, the block level assertions shall be included in the system level verification by default. The system interfaces provide a more realistic and unexpected set of input conditions and block level assertions must be able to react to them correctly.

- The verification environment should provide the facility to turn on block level assertions if there are any failures. For example, in our sample system, if a failure occurs during a transaction between master 1 and target 0, then the system level simulation should be re-run by including the block level SVA checks written for master 1 and target 0.

- At the system level, a new set of assertions should be written that verifies the connectivity of the system. More focus should be on the interface rules rather than the internal block details.

2.3.1 SVA Checks for system level verification

The following set of checks can be written for the system level verification based on the connectivity and protocol of the system.

Ss_shk1: Only one "trdy" signal can be asserted at any given point. In other words, only one target device can participate in a transaction at any given time.

```
property p_target;
 @(posedge clk) not (!trdy[0] && !trdy[1]);
endproperty
```

Ss_chk2: Only one set of "frame" and "irdy" signals can be asserted at any given clock cycle. In other words, only one master device can participate in a transaction at any give time.

```
property p_frame;
 @(posedge clk)
     $countones({frame1, frame2, frame3}) >1;
endproperty

property p_irdy;
 @(posedge clk)
     $countones({irdy1, irdy2, irdy3}) >1;
endproperty
```

Ss_chk3: Only one "gnt" signal shall be asserted at any given time. In other words, the arbiter can provide access for only one master at a time to pursue a transaction.

```
property p_gnt;
 @(posedge clk)
       $countones({gnt1, gnt2, gnt3}) > 1;
endproperty
```

Ss_chk4: Only one "rw" signal shall be active at any given clock cycle, the other "rw" signals should be tri-stated ("rw" signal is the MSB of the masters data output bus).

```
property p_rw;
 @(posedge clk)
 ($isunknown(rw1)     &&     $isunknown(rw2)     &&
 $isunknown(rw3) ) ||
 ((rw1==1'b1 || rw1==1'b0) && $isunknown (rw2)
 && $isunknown(rw3)) ||
 ((rw2==1'b1 || rw2==1'b0) && $isunknown (rw1)
 && $isunknown(rw3)) ||
```

```
  ((rw3==1'b1  || rw3==1'b0)  && $isunknown (rw2)
  && $isunknown(rw2));
endproperty
```

Ss_chk5: Only one "rsel" signal shall be active at any given clock cycle, the other "rsel" signals should be tri-stated.

```
property p_rsel;
  @(posedge clk)
$isunknown(rsel1)   &&   $isunknown(rsel2)   &&
$isunknown(rsel3) ) ||
  ((rsel1==1'b1  ||  rsel1==1'b0)  &&  $isunknown
(rsel2) && $isunknown(rsel3)) ||
  ((rsel2==1'b1  ||  rsel2==1'b0)  &&  $isunknown
(rsel1) && $isunknown(rsel3)) ||
  ((rsel3==1'b1  ||  rsel3==1'b0)  &&  $isunknown
(rsel2) && $isunknown(rsel1));
endproperty
```

Ss_chk6: Upon a valid request by a master, a valid "gnt" should arrive within 2 to 5 clock cycles.

```
assign req = !req1 || !req2 || !req3;
assign gnt = !gnt1 || !gnt2 || !gnt3;

property p_req_gnt_w;
  @(posedge clk)
        $rose (req) |-> ##[2:5] $rose(gnt);
endproperty
```

Ss_chk7: At any given clock, if the "frame" and "irdy" signal of a master are asserted, then the relevant "trdy" signal should be asserted after 3 clock cycles.

```
assign frame_ = !frame1 || !frame2 || !frame3;
assign irdy_ = !irdy1 || !irdy2 || !irdy3;

property p_start_frame;
  @(posedge clk)
      $rose (frame_ && irdy_) |->##3 $rose(trdy_);
endproperty
```

Ss_chk8: At any given clock, if the "frame" and "irdy" signals of the master are de-asserted, then the relevant "trdy" signal should be de-asserted after 2 clock cycles.

```
assign trdyp = trdy[1] && trdy[0];

property p_end_frame;
 @(posedge clk)
       $rose (frame && irdy) |->##2 $rose(trdyp);
endproperty
```

Ss_chk9: If there is no valid transaction at any given clock, then the bus "data" and "datao" should be tri-stated.

```
property p_bus_not_in_use;
 @(posedge clk)
  trdyp |->
       ($isunknown(data) && $isunknown(datao));
endproperty

a_target : assert property(p_target);
a_frame: assert property(p_frame);
a_irdy: assert property(p_irdy);
a_rsel: assert property(p_rsel);
a_rw: assert property(p_rw);
a_gnt: assert property(p_gnt);
a_req_gnt_w : assert property(p_req_gnt_w);
a_start_frame: assert property(p_start_frame);
a_end_frame: assert property(p_end_frame);
a_bus_in_use: assert property(p_bus_not_in_use);

c_target : cover property(p_target);
c_frame: cover property(p_frame);
c_irdy: cover property(p_irdy);
c_rsel: cover property(p_rsel);
c_rw: cover property(p_rw);
c_gnt: cover property(p_gnt);
c_req_gnt_w : cover property(p_req_gnt_w);
c_start_frame: cover property(p_start_frame);
c_end_frame: cover property(p_end_frame);
c_bus_in_use: cover property(p_bus_not_in_use);
```

During the system level simulation, the top-level module should be configured with the parameter settings such that all block level assertions are turned off. In our sample system, since each design block has a parameter that allows including its relevant SVA checks on a need basis, we can configure the top module for system level run easily as shown below.

```
Module top(..,..,     );

// port declarations

master
#(.master_sva(1'b0),  .master_sva_severity(1'b0))
u1 (ask[2], clk, req1, gnt1, frame1, irdy1, trdy,
data1, rsel1, datao);

master
#(.master_sva(1'b0),  .master_sva_severity(1'b0))
u2 (ask[1], clk, req2, gnt2, frame2, irdy2, trdy,
data2, rsel2, datao);

master
#(.master_sva(1'b0),  .master_sva_severity(1'b0))
u3 (ask[0], clk, req3, gnt3, frame3, irdy3, trdy,
data3, rsel3, datao);

arbiter
#(.arb_sva(1'b0), .arb_sva_severity(1'b0))
u4 (clk, reset, frame, irdy, req1, req2, req3,
gnt1, gnt2, gnt3);

glue
#(.glue_sva(1'b0), .glue_sva_severity(1'b0))
u5 (clk, frame1, irdy1, frame2, irdy2, frame3,
irdy3, trdy, rsel1, rsel2, rsel3, data1, data2,
data3, sel, data, dataout1, dataout2, datao);

target
#(.target_sva(1'b0),  .target_sva_severity(1'b0))
u6 (clk, reset, sel[1], trdy[1], data, dataout1);

target
#(.target_sva(1'b0),  .target_sva_severity(1'b0))
u7 (clk, reset, sel[0], trdy[0], data, dataout2);
```

```
endmodule
```

Note that when each design block is instantiated, the parameter values are passed. The first parameter "*_sva" is set to 0 in all the individual instantiations, which indicates that the block level assertions will not be included. Now, the system level simulations can be run only with the system level checks.

Let us assume that there are failures on "Ss_chk6" during the system level simulation. This check looks for interface failures between the masters and the arbiter module. To debug the errors, the simulation can be re-run by including the block level checks relevant to the masters and the arbiter. The top modules configuration for such a run is shown below:

```
Module top(..,..,    );

// port declarations

master
#(.master_sva(1'b1),   .master_sva_severity(1'b0))
u1 (ask[2], clk, req1, gnt1, frame1, irdy1, trdy,
data1, rsel1, datao);

master
#(.master_sva(1'b1),   .master_sva_severity(1'b0))
u2 (ask[1], clk, req2, gnt2, frame2, irdy2, trdy,
data2, rsel2, datao);

master
#(.master_sva(1'b1),   .master_sva_severity(1'b0))
u3 (ask[0], clk, req3, gnt3, frame3, irdy3, trdy,
data3, rsel3, datao);

arbiter
#(.arb_sva(1'b1), .arb_sva_severity(1'b0))
u4 (clk, reset, frame, irdy, req1, req2, req3,
gnt1, gnt2, gnt3);

glue
#(.glue_sva(1'b0), .glue_sva_severity(1'b0))
u5 (clk, frame1, irdy1, frame2, irdy2, frame3,
irdy3, trdy, rsel1, rsel2, rsel3, data1, data2,
data3, sel, data, dataout1, dataout2, datao);
```

```
target
 #(.target_sva(1'b0),  .target_sva_severity(1'b0))
u6 (clk, reset, sel[1], trdy[1], data, dataout1);

target
#(.target_sva(1'b0),   .target_sva_severity(1'b0))
u7 (clk, reset, sel[0], trdy[0], data, dataout2);

endmodule
```

Note that the parameter "master_sva" and "arb_sva" are set to 1 in this configuration. In the basic design blocks, SVA checks could also be included conditionally using the "`ifdef - `endif" construct. By conditionally compiling the SVA code, the user can either have the checks on all instances of the module or on none of the instances of the module. The disadvantage with this methodology is that, it is a global control mechanism. By using parameters, this disadvantage can be overcome and the user gets more flexibility in choosing the block level checks needed for a particular simulation run.

2.4 Functional coverage

The system level checks written so far look for specific protocol violations, if any. By making sure that these checks executed at least once in the simulation, the confidence level on the functionality of the system increases tremendously. The other aspect of functional coverage is covering all possible scenarios of system functionality during simulation from the testbench perspective. ***The scenarios to be covered during a simulation should be part of the test plan.***

The SVA checks written for dynamic simulation are only as good as the input stimulus. If the input vectors do not force the system to execute certain scenarios, then those remain untested. A lot of testbenches use random techniques to generate input stimulus vectors. A very common approach is to run a pre-determined number of transactions and measure coverage on certain scenarios. By constraining the random generation of input stimulus, the scenarios can be covered more efficiently. ***The key is to get the maximum functional coverage in a minimum number of cycles.*** The coverage information collected from SVA can be used effectively to create reactive verification environments.

2.4.1 Coverage plan for the sample system

The sample system discussed in this chapter has a lot of key functionality that should be covered as part of the functional verification.

2.4.1.1 Request Scenario

"All possible request scenarios should be covered"

There are three masters that can ask for access at any given time. This means that there are 7 possible combinations of the master "req" signals as shown in Table 2-1.

Table 2-1. Master request scenarios

Req1	Req2	Req3
0	1	1
1	0	1
1	1	0
0	0	1
1	0	0
0	1	0
0	0	0

A 0 in the table indicates that the master is requesting for the bus. The testbench should create all these possible input combinations during simulation.

The following code example shows how functional coverage data can be used to control the simulation environment. Property definitions for all 7 possible request combinations should be created as follows.

```
property p_req1; // master 1 requesting
  @(posedge clk) $fell (req1) && req2 && req3;
endproperty

property p_req2; // master 2 requesting
  @(posedge clk) $fell (req2) && req1 && req3;
endproperty

property p_req3; // master 3 requesting
  @(posedge clk) $fell (req3) && req1 && req2;
endproperty
```

```
property p_req12; // master 1&2 requesting
  @(posedge clk)
  $fell (req1) && $fell(req2)&& req3;
endproperty

property p_req23; // master 2&3 rtequesting
  @(posedge clk)
  $fell (req2) && $fell(req3) && req1;
endproperty

property p_req31; // master 1&3 requesting
  @(posedge clk)
  $fell (req3) && $fell(req1) && req2;
endproperty

property p_req123; // master 1&2&3 requesting
  @(posedge clk)
  $fell (req1) && $fell(req2) && $fell(req3);
endproperty
```

Each property should have a cover statement associated with it as shown below. The action block of the cover statement can be used to update register flags. In this case, every time the property is covered, a local register count is incremented. In the same clock, we check if the counter has reached a value of 3. If so, then the flag associated to that property is asserted. In other words, it is expected that each request combination occurs three times during simulation and if and when it happens, a flag associated with that specific request combination will be asserted.

```
c_req1: cover property(p_req1)
   begin
     creq1++;
     if(creq1 == 3) creq1_flag = 1'b1;
   end

c_req2: cover property(p_req2)
   begin
     creq2++;
     if(creq2 == 3) creq2_flag = 1'b1;
   end

c_req3: cover property(p_req3)
   begin
```

```
      creq3++;
      if(creq3 == 3) creq3_flag = 1'b1;
    end
c_req12: cover property(p_req12)
    begin
      creq12++;
      if(creq12 == 3) creq12_flag = 1'b1;
    end

c_req23: cover property(p_req23)
    begin
      creq23++;
      if(creq23 == 3) creq23_flag = 1'b1;
    end

c_req31: cover property(p_req31)
    begin
      creq31++;
      if(creq31 == 3) creq31_flag = 1'b1;
    end

c_req123: cover property(p_req123)
    begin
      creq123++;
      if(creq123 == 3) creq123_flag = 1'b1;
    end
```

This coverage information can be used effectively to control the simulation environment. In a random testbench for the sample system, a pre-determined number of transactions could be performed one after the other. The simulation will finish when all transactions are completed. The following code shows how the functional coverage information can be used to terminate the simulation.

```
always@(posedge clk)
begin

if(creq1_flag && creq2_flag && creq3_flag &&
creq12_flag && creq23_flag && creq31_flag &&
creq123_flag)

begin
```

```
    $display("FC: All possible request scenarios
    covered 3 times each\n");
    $finish();

  end
  end
```

With this piece of code, there are two ways to terminate a simulation:

1. Run the pre-determined number of transactions randomly and exit.
2. Exit if all possible request scenarios are covered three times each.

Whichever occurs first will terminate the simulation.

2.4.1.2 Master to Target transactions

"Every master device should perform both a read and a write transaction with every target device"

There are 3 master devices and 2 target devices in the system. This creates 12 possible scenarios as shown in Table 2-2. Property definitions for all 12 possible transaction combinations should be created as follows.

Table 2-2. Master to target transactions

Master	Target	Transaction
M1	T1	Read
M1	T1	Write
M1	T0	Read
M1	T0	Write
M2	T1	Read
M2	T1	Write
M2	T0	Read
M2	T0	Write
M3	T1	Read
M3	T1	Write
M3	T0	Read
M3	T0	Write

```
property p_m1t1r;
// master1 reading from target 1
  @(posedge clk)
  $fell (frame1 && irdy1) |->
```

```
      ##3 ($fell (trdy[1])) ##3 !data[8];
endproperty

property p_m1t1w;
// master 1 writing to target 1
 @(posedge clk)
 $fell (frame1 && irdy1) |->
      ##3 ($fell (trdy[1])) ##3 data[8];
endproperty

property p_m1t0r;
// master 1 reading from target 0
 @(posedge clk)
 $fell (frame1 && irdy1) |->
      ##3 ($fell (trdy[0])) ##3 !data[8];
endproperty

property p_m1t0w;
// master 1 writing to target 0
 @(posedge clk)
 $fell(frame1 && irdy1) |->
      ##3 ($fell(trdy[0])) ##3 data[8];
endproperty

property p_m2t1r;
// master 2 reading from target 1
 @(posedge clk)
 $fell (frame2 && irdy2) |->
      ##3 ($fell(trdy[1])) ##3 !data[8];
endproperty

property p_m2t1w;
// master 2 writing to target 1
 @(posedge clk)
 $fell (frame2 && irdy2) |->
      ##3 ($fell (trdy[1])) ##3 data[8];
endproperty

property p_m2t0r;
// master 2 reading from target 0
 @(posedge clk)
 $fell (frame2 && irdy2) |->
      ##3 ($felltrdy[0])) ##3 !data[8];
```

```
endproperty

property p_m2t0w;
// master 2 writing to target 0
  @(posedge clk)
  $fell (frame2 && irdy2) |->
         ##3 ($fell (trdy[0])) ##3 data[8];
endproperty

property p_m3t1r;
// master 3 reading from target 1
  @(posedge clk)
  $fell (frame3 && irdy3) |->
         ##3 ($fell (trdy[1])) ##3 !data[8];
endproperty

property p_m3t1w;
// master 3 writing to target 1
  @(posedge clk)
  $fell (frame3 && irdy3) |->
         ##3 ($fell (trdy[1])) ##3 data[8];
endproperty

property p_m3t0r;
// master 3 reading from target 0
  @(posedge clk)
  $fell (frame3 && irdy3) |->
         ##3 ($fell (trdy[0])) ##3 !data[8];
endproperty

property p_m3t0w;
// master 3 writing to target 0
  @(posedge clk)
  $fell (frame3 && irdy3) |->
         ##3 ($fell (trdy[0])) ##3 data[8];
endproperty
```

Each property should have a cover statement associated with it as shown below. The same technique used in Section 2.4.1.1 is used to keep count of the number of occurrences of the scenario.

```
c_m1t1r: cover property(p_m1t1r)
   begin
```

```
      m1_t1_r++;
      if(m1_t1_r == 3) m1_t1_r_flag = 1'b1;
    end

c_m1t1w: cover property(p_m1t1w)
    begin
      m1_t1_w++;
      if(m1_t1_w == 3) m1_t1_w_flag = 1'b1;
    end

c_m1t0r: cover property(p_m1t0r)
    begin
      m1_t0_r++;
      if(m1_t0_r == 3) m1_t0_r_flag = 1'b1;
    end

c_m1t0w: cover property(p_m1t0w)
    begin
      m1_t0_w++;
      if(m1_t0_w == 3) m1_t0_w_flag = 1'b1;
    end

c_m2t1r: cover property(p_m2t1r)
    begin
      m2_t1_r++;
      if(m2_t1_r == 3) m2_t1_r_flag = 1'b1;
    end

c_m2t1w: cover property(p_m2t1w)
    begin
      m2_t1_w++;
      if(m2_t1_w == 3) m2_t1_w_flag = 1'b1;
    end

c_m2t0r: cover property(p_m2t0r)
    begin
      m2_t0_r++;
      if(m2_t0_r == 3) m2_t0_r_flag = 1'b1;
    end

c_m2t0w: cover property(p_m2t0w)
    begin
      m2_t0_w++;
```

```
      if(m2_t0_w == 3) m2_t0_w_flag = 1'b1;
    end

c_m3t1r: cover property(p_m3t1r)
    begin
      m3_t1_r++;
      if(m3_t1_r == 3) m3_t1_r_flag = 1'b1;
    end

c_m3t1w: cover property(p_m3t1w)
    begin
      m3_t1_w++;
      if(m3_t1_w == 3) m3_t1_w_flag = 1'b1;
    end

c_m3t0r: cover property(p_m3t0r)
    begin
      m3_t0_r++;
      if(m3_t0_r == 3) m3_t0_r_flag = 1'b1;
    end

c_m3t0w: cover property(p_m3t0w)
    begin
      m3_t0_w++;
      if(m3_t0_w == 3) m3_t0_w_flag = 1'b1;
    end
```

This coverage information from both Sections 2.4.1.1 and 2.4.1.2 can be used effectively to control the simulation environment. With the piece of code shown below, there are two ways to terminate a simulation:

1. Run a pre-determined number of transactions randomly and exit.
2. If all possible request scenarios are covered three times and if all possible "master to target" transactions are covered three times, then exit the simulation.

Whichever occurs first will terminate the simulation.

```
always@(posedge clk)
begin

if(creq1_flag && creq2_flag && creq3_flag &&
creq12_flag && creq23_flag && creq31_flag &&
```

```
creq123_flag && m1_t1_r_flag && m1_t1_w_flag &&
m1_t0_r_flag && m1_t0_w_flag && m2_t1_r_flag &&
m2_t1_w_flag && m2_t0_r_flag && m2_t0_w_flag &&
m3_t1_r_flag && m3_t1_w_flag && m3_t0_r_flag &&
m3_t0_w_flag)

 begin

   $display("FC: All possible request scenarios
   covered 3 times\n");

   $display("FC: All possible transactions covered
   3 times\n");

   $finish();

 end
 end
```

2.4.1.3 Advanced coverage options

There is another data point that can be used to measure the functional coverage of the system.

"Every target memory location should be written to and read from at least once by each master"

This information requires exhaustive testing. Every address space in the target device should be monitored for usage by each master device. SVA is not always the choice for performing functional coverage. ***Functional coverage that involves exhaustive test plan coverage points can be done more efficiently with a testbench language that supports object oriented programming constructs.*** Such exhaustive functional coverage points should be used while running long regression runs.

2.4.2 Functional coverage summary

Functional coverage measurement guarantees testing of all required scenarios. The measure can be used effectively for controlling simulation environments. One method is to terminate simulation upon achieving the functional coverage goals. In the sample system, the following results were observed:

- Default number of random transactions set in the testbench was 500.
- Terminating the simulation based on the request scenarios shown in Section 2.4.1.1 took 46 transactions.
- Terminating the simulation based on the request scenarios shown in Section 2.4.1.1 and the "master to target" transactions shown in Section 2.4.1.2 took 63 transactions.

The functional coverage data obtained can also be used to re-direct the testbench dynamically. In random testbenches, constraints are used to control the type of transactions generated. These constraints are assigned certain weights for the random distribution in the beginning of a simulation. Based on the functional coverage information obtained during the simulation, these weights can be adjusted dynamically to achieve the functional coverage goal quickly.

2.5 SVA for transaction log creation

SVA can be used to create excellent log files. The SVA checkers snoop for any design property violation during simulation. The same checkers can be called monitors if they log the information that they are snooping. In a complex system, it really helps to create a chronological log of the transactions. In our sample system, creating a log of all the read and write transactions, between whom these happened and at what time will be a great debugging asset.

SVA has the option to use a lot of the Verilog like capabilities within the scope of the checker. The action block of each checker or cover statement can be used efficiently to create log files. While displaying information upon the success of an assert or a cover statement is one way to create log files, another way is to call a task or a function. The calling of a task or a function expands the capabilities of the SVA checker. Apart from displaying information within the task, data checking can also be done effectively. The following code shows how a chronological transaction log is created for the sample system.

```
// open a file to document transactions

integer h_mt;
initial
begin
```

```
 h_mt = $fopen("mt.dat");
end

// calling task for documentation

`ifdef slv_doc

c_m1t1w_doc:
   cover property(p_m1t1w) master_xaction(1,1);
c_m1t1r_doc:
   cover property(p_m1t1r) master_xaction(1,1);
c_m1t2w_doc:
   cover property(p_m1t0w) master_xaction(1,0);
c_m1t2r_doc:
   cover property(p_m1t0r) master_xaction(1,0);
c_m2t1w_doc:
   cover property(p_m2t1w) master_xaction(2,1);
c_m2t1r_doc:
   cover property(p_m2t1r) master_xaction(2,1);
c_m2t2w_doc:
   cover property(p_m2t0w) master_xaction(2,0);
c_m2t2r_doc:
   cover property(p_m2t0r) master_xaction(2,0);
c_m3t1w_doc:
   cover property(p_m3t1w) master_xaction(3,1);
c_m3t1r_doc:
   cover property(p_m3t1r) master_xaction(3,1);
c_m3t2w_doc:
   cover property(p_m3t0w) master_xaction(3,0);
c_m3t2r_doc:
   cover property(p_m3t0r) master_xaction(3,0);

`endif

task master_xaction(
 input int m_identity, input int t_identity);

integer i;

begin

if(data[8])
begin
```

```
for(i=0; i<8; i++)
begin

$fwrite(h_mt,"WRITE:
Master %0d writing to Target %0d = %0d at
%0t\n",m_identity, t_identity, data[7:0],
$time);

@(posedge clk);
end
end

if(!data[8])
begin
@(posedge clk);
for(i=0; i<8; i++)
begin
$fwrite(h_mt,"READ:
Master %0d reading from Target %0d = %0d at
%0t\n", m_identity, t_identity, datao, $time);

@(posedge clk);
end
end

end

endtask
```

The properties defined for functional coverage in Section 2.4.1.2 are reused for creating transaction logs. If the cover statement succeeds, a task called "master_xaction" is called. The task expects two input arguments, one identifying the master and the other identifying the target device. By sending these arguments, a generic task can be written to log the transactions accurately.

The transactions are logged into a separate file called "mt.dat." A $fopen statement is used to open this file at the beginning of the simulation. Once the task is called, the task executes either the read block of the code or the write block of the code. Since our sample system does burst read or write in sets of 8 bytes, a "for" loop is used within the task. The loop goes around eight times and each time the relevant read or write data is logged into the file "mt.dat" using a $fwrite statement. A part of the log created for the sample system using this code is shown below.

```
WRITE: Master 1 writing to Target 1 = 72 at 775
WRITE: Master 1 writing to Target 1 = 77 at 825
WRITE: Master 1 writing to Target 1 = 95 at 875
WRITE: Master 1 writing to Target 1 = 37 at 925
WRITE: Master 1 writing to Target 1 = 216 at 975
WRITE: Master 1 writing to Target 1 = 184 at 1025
WRITE: Master 1 writing to Target 1 = 198 at 1075
WRITE: Master 1 writing to Target 1 = 182 at 1125
READ: Master 3 reading from Target 1 = 72 at 1725
READ: Master 3 reading from Target 1 = 77 at 1775
READ: Master 3 reading from Target 1 = 95 at 1825
READ: Master 3 reading from Target 1 = 37 at 1875
READ:Master 3 reading from Target 1 = 216 at 1925
READ:Master 3 reading from Target 1 = 184 at 1975
READ:Master 3 reading from Target 1 = 198 at 2025
READ:Master 3 reading from Target 1 = 182 at 2075
```

The transaction logs can be made a lot more fancy and debug friendly depending on the user's application. Note that this code is included within the `ifdef - `endif block. This kind of a detailed transaction log might not be needed during long regressions and hence should have the provision to be included conditionally.

2.6 SVA for FPGA Prototyping

A variety of advanced verification methodologies exist today that can help find bugs quickly. Constrained random testbenches and assertions are an important piece in these methodologies. It is very common to write thousands of tests to make sure that all possible functionality has been tested correctly. While most of the bugs are found in the RTL verification, it is still very common to find functional bugs during the verification of implemented gates. Simulating gates has always been a performance bottleneck and will always be. Running all the tests developed during RTL verification on gates is not very practical. Gate level simulation is extremely slow and more and more verification teams are depending on other verification methodologies such as formal verification, FPGA prototyping, etc. as shown in Figure 2-14. By running the verification on the actual silicon, the verification process can be accelerated significantly. This allows running the regression suites developed for RTL exhaustively on actual silicon.

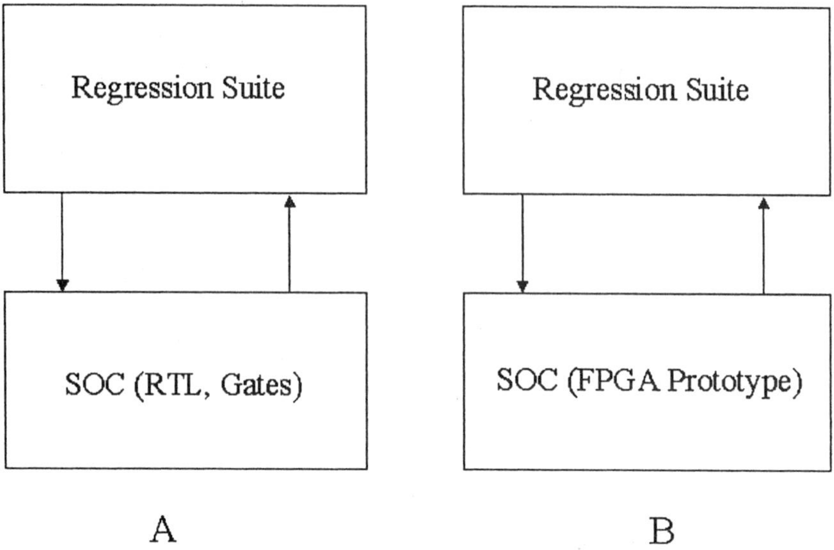

Figure 2-14. FPGA Prototyping

One major challenge in running tests on actual silicon prototype is debugging. SVA can help in this area significantly. By synthesizing the checkers along with the design, the debug process can be made a little easier. The checkers are written against the functional specification and having them monitor the design in real silicon adds great value. The design needs to be altered slightly to accommodate these assertions. If an assertion fails, it has to be notified to the external world using an output port. The output ports can be updated with the results, using the action block of the assertions. In most real-time testing, breakpoints can be set on these output ports and upon a failure on one of these debug ports, the verification can be stopped for further analysis. The master device used in the sample system is shown in Figure 2-2. This contains only the default ports relevant to the design. The sample Verilog code for the master device is shown below.

```
module master (ask_for_it, clk, req, gnt, frame,
irdy, trdy, data_c, r_sel, data_o);

input clk, gnt, ask_for_it;
input [1:0] trdy;
output req, frame, irdy, r_sel;
```

```
output [8:0] data_c;
input [7:0] data_o;

parameter master_sva = 1'b1;
parameter master_sva_severity = 1'b1;

// functional description of master

// Block level SVA checks

endmodule
```

The block level assertions should be made part of the design to help in FPGA prototyping. Each block level assertion should be associated with a debug output port. The debug output port should be asserted if the assertion fails. The following code description shows how this can be achieved.

```
module master (ask_for_it, clk, req, gnt, frame,
irdy,     trdy,     data_c,     r_sel,     data_o,
a_master_start1_flag, a_master_start2_flag,
a_master_stop1_flag, a_master_stop2_flag,
a_master_data1_flag, a_master_data2_flag,
a_master_datao1_flag, a_master_datao2_flag);

input clk, gnt, ask_for_it;
input [1:0] trdy;
output req, frame, irdy, r_sel;
output [8:0] data_c;
input [7:0] data_o;

// debug pins for FPGA prototyping
output a_master_start1_flag;
output a_master_start2_flag;
output a_master_stop1_flag;
output a_master_stop2_flag;
output a_master_data1_flag;
output a_master_data2_flag;
output a_master_datao1_flag;
output a_master_datao2_flag;

parameter master_sva = 1'b1;
parameter master_sva_severity = 1'b1;
```

```
// functional description of master

// Block level checks for prototype debugging

`ifdef master_debug

d_a_master_start1:
   assert property(p_master_start1)
     else
       a_master_start1_flag = 1'b1;
d_a_master_start2:
     assert property(p_master_start2)
     else
       a_master_start2_flag = 1'b1;
d_a_master_stop1:
     assert property(p_master_stop1)
     else
       a_master_stop1_flag = 1'b1;
d_a_master_stop2:
     assert property(p_master_stop2)
     else
       a_master_stop2_flag = 1'b1;
d_a_master_data1:
     assert property(p_master_data1)
     else
       a_master_data1_flag = 1'b1;
d_a_master_data2:
     assert property(p_master_data2)
     else
       a_master_data2_flag = 1'b1;
d_a_master_datao1:
     assert property(p_master_datao1)
     else
       a_master_datao1_flag = 1'b1;
d_a_master_datao2:
     assert property(p_master_datao2)
     else
       a_master_datao2_flag = 1'b1;

`endif

endmodule
```

Note that the respective output port flags will be asserted upon a failure. Since these assertions are concurrent, they will look for a valid start on every clock edge. If the silicon testing mechanism does not provide a way to set breakpoints on an assertion failure, then it is required that the failure be latched. Otherwise, the failure notification can be lost if the assertion succeeds in future clock cycles.

2.7 Summary on SVA simulation methodologies

- The addition of SVA to testbench environment makes dynamic simulation more productive.
- The designers are very familiar with the internal functionality of the design and hence, they should in-line SVA checkers in their respective design blocks.
- The verification engineer, who integrates and verifies the system, should add system level assertions that thoroughly verify the interface protocol.
- The verification engineer should be able to control/configure the block level assertions from his verification environment (He should be able to turn the assertions on and off on a need basis).
- Functional coverage metrics can be collected with little effort using SVA. This information should be used effectively to create reactive testbenches.
- SVA can be used to create informative log files since they are monitoring the design protocols throughout the simulation.
- By writing SVA checkers that follow synthesis coding guidelines, they can be made part of the net-list and used to debug prototyping/emulation failures.

Chapter 3

SVA FOR FINITE STATE MACHINES

FSM is the main control block in any design. It helps the design progress from state to state in an orderly manner by generating the respective control signals. Another way to generate control signals is by combining counters and glue logic. But it lacks good design structure and is also difficult to debug. An FSM provides great hardware infrastructure for control signals and also debugging capabilities since each state of the design is usually well defined.

There are two types of FSMs:

Moore State machine – The Moore FSM outputs are the function of the present state only.

Mealy State machine – One or more of the Mealy FSM outputs are a function of the present state and one or more of the inputs.

Different types of coding styles are used to describe the states of an FSM. The most popular coding style is the one-hot coding, wherein a one-bit register represents each state. This proves to be the fastest architecture. If the FSM has too many states, then one-hot coding will produce a rather big hardware. In these cases binary encoding is preferred. Another kind of encoding used commonly to describe an FSM is the gray coding.

An FSM controls the functionality of the entire design and hence should be verified thoroughly. The most common type of check is to make sure that the state transitions are occurring correctly without violating any timing requirement. SVA can be used effectively to do such checks.

3.1 Sample Design – FSM1

In this section, we analyze a simple linear FSM, which is more like a shift counter. The FSM produces control signals for the design in a linear sequential fashion and hence can be verified easily with SVA checks.

3.1.1 Functional description of FSM1

There are 16 states in the FSM. They are coded as follows:

IDLE = 16'd1
GEN_BLK_ADDR = 16'd2
WAIT6 = 16'd4
NEXT_BLK = 16'd8
WAIT0 = 16'd16
CNT1 = 16'd32
WAIT1 = 16'd64
CNT2 = 16'd128,
WAIT2 = 16'd256
CNT3 = 16'd512
WAIT3 = 16'd1024
CNT4 = 16'd2048
WAIT4 = 16'd4096
CNT5 = 16'd8192
WAIT5 = 16'd16384
CNT6 = 16'd32768

The FSM is coded with a one-hot coding style. Figure 3-1 shows the bubble diagram of FSM1:

- The FSM moves to the IDLE state upon reset and waits there for a valid "get_data" signal.
- Once a valid "get_data" signal is obtained, the FSM moves to the GEN_BLK_ADDR state. The FSM stays in this state until it finishes generating 64 read addresses (an internal counter keeps track of 64 clock cycles).
- After 64 clock cycles, it moves to the WAIT0 state. From this point, the FSM keeps moving to the next state on every clock cycle.
- The CNT* states are the ones where the output control signals for the rest of the design are generated.

- The WAIT* states are used to create a 2 cycle gap between the generation of the control signals (latch_en, dp1_en, dp2_en, dp3_en, dp4_en, wr).
- Once the FSM moves to the NEXT_BLK state it has to decide which way to go.

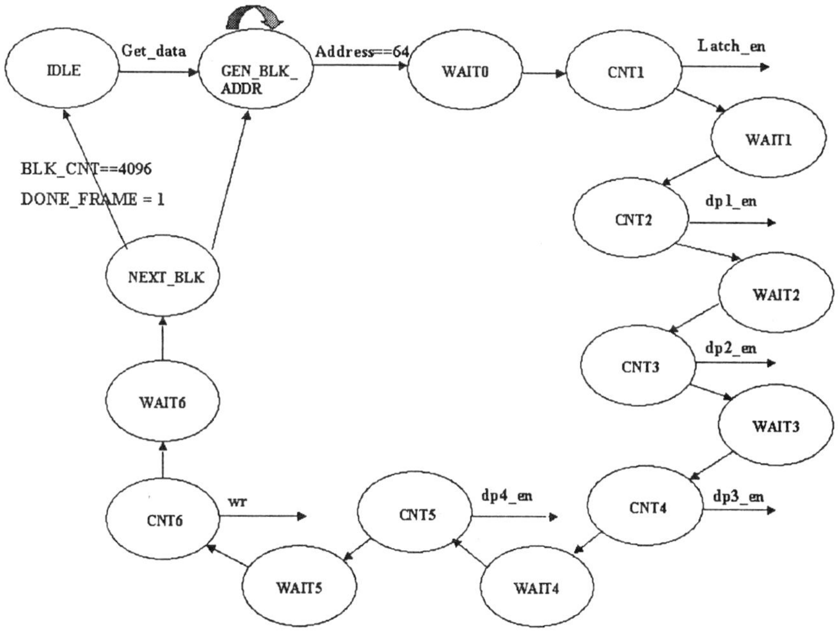

Figure 3-1. Bubble diagram for FSM1

- If the internal register "blk_cnt" has reached the value of 4096, the FSM goes to the IDLE state. This indicates that the entire data frame has been processed and the design is waiting for a new frame. When a new "get_data" signal arrives, the FSM goes through the same state transitions again starting from GEN_BLK_ADDR.
- While in state NEXT_BLK, if the internal register "blk_cnt" has not reached the value of 4096, the FSM will go back to the GEN_BLK_ADDR state. It waits for the generation of 64 new addresses and then moves over to the CNT* states to generate the control signals again.

Example 3.1 FSM1 sample code

```
module fsm (get_data, reset_, clk, rd, rd_addr,
data, done_frame, latch_en, sipo_en, dp1_en,
dp2_en, dp3_en, dp4_en, wr);

  input get_data;
  input reset_;
  input clk;
  input [7:0] data;

  output rd;
  output logic sipo_en, latch_en;
  output logic dp1_en, dp2_en, dp3_en, dp4_en, wr;
  output logic done_frame;
  output [17:0] rd_addr;

  logic [5:0] addr_cnt;
  logic [11:0] blk_cnt;
  logic [3:0]  pipeline_cnt;
  logic rd;
  logic [17:0] rd_addr;
  logic enable_cnt, enable_dly_cnt, enable_blk_cnt;

  assign done_frame = (blk_cnt == 4095);
  assign sipo_en = rd;

   enum bit[15:0] {IDLE = 16'd1,
       GEN_BLK_ADDR = 16'd2,
       DLY = 16'd4,
       NEXT_BLK = 16'd8,
       WAIT0 = 16'd16,
       CNT1 = 16'd32,
       WAIT1 = 16'd64,
       CNT2 = 16'd128,
       WAIT2 = 16'd256,
       CNT3 = 16'd512,
       WAIT3 = 16'd1024,
       CNT4 = 16'd2048,
       WAIT4 = 16'd4096,
       CNT5 = 16'd8192,
       WAIT5 = 16'd16384,
       CNT6 = 16'd32768} n_state, c_state;
```

```
// assign the different control signals

assign latch_en = (c_state == CNT1);
assign dp1_en = (c_state == CNT2);
assign dp2_en = (c_state == CNT3);
assign dp3_en = (c_state == CNT4);
assign dp4_en = (c_state == CNT5);
assign wr = (c_state == CNT6);

// 64bit counter to generate read address

always_ff @(posedge clk)
if (!reset_ || !enable_cnt)
 addr_cnt <= 0;
else if (enable_cnt)
 addr_cnt <= addr_cnt + 1;
else
 addr_cnt <= addr_cnt;

// 4096 bit counter

always_ff @(posedge clk)
if (!reset_)
 blk_cnt <= 0;
else if ((c_state == NEXT_BLK) && enable_blk_cnt)
 blk_cnt <= blk_cnt + 1;
else
 blk_cnt <= blk_cnt;

always_ff @(posedge clk)
  if (!reset_)
    c_state <= IDLE;
     else
    c_state <= n_state;

always @(*)
 begin
 rd <= 0;
 enable_cnt <= 0;
 //enable_dly_cnt <= 0;
 case (c_state)
   IDLE:  begin
        enable_blk_cnt <= 0;
```

```
        if(get_data)
            n_state <= GEN_BLK_ADDR;
        else
            n_state <= IDLE;
          end

    GEN_BLK_ADDR: begin
          enable_cnt <= 1;
          rd <= 1;
          rd_addr <= {blk_cnt, addr_cnt};
          if (addr_cnt == 63) begin
              //enable_dly_cnt <= 1;
              n_state <= WAIT0;
            end
          else begin
              n_state <= GEN_BLK_ADDR;
              //pipeline_cnt <= 0;
            end
          end

    WAIT0: n_state <= CNT1;
    CNT1: n_state <= WAIT1;
    WAIT1: n_state <= CNT2;
    CNT2: n_state <= WAIT2;
    WAIT2: n_state <= CNT3;
    CNT3: n_state <= WAIT3;
    WAIT3: n_state <= CNT4;
    CNT4: n_state <= WAIT4;
    WAIT4: n_state <= CNT5;
    CNT5: n_state <= WAIT5;
    WAIT5: n_state <= CNT6;
    CNT6: n_state <= DLY;

    DLY: begin
        enable_blk_cnt <= 1;
        n_state <= NEXT_BLK;
      end


NEXT_BLK: begin
          enable_blk_cnt<=1;
          if (blk_cnt == 4095)
              n_state <= IDLE;
```

```
            else
                n_state <= GEN_BLK_ADDR;
            end
    endcase
  end

  endmodule
```

The state transition from GEN_BLK_ADDR to the CNT*/WAIT* states is shown in Figure 3-2.

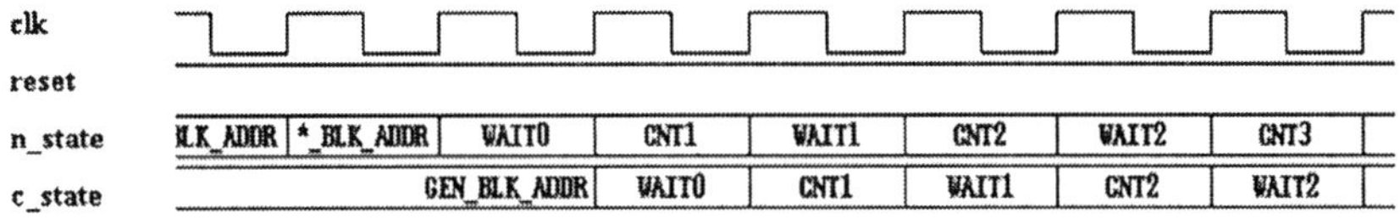

Figure 3-2. Waveform A for FSM1

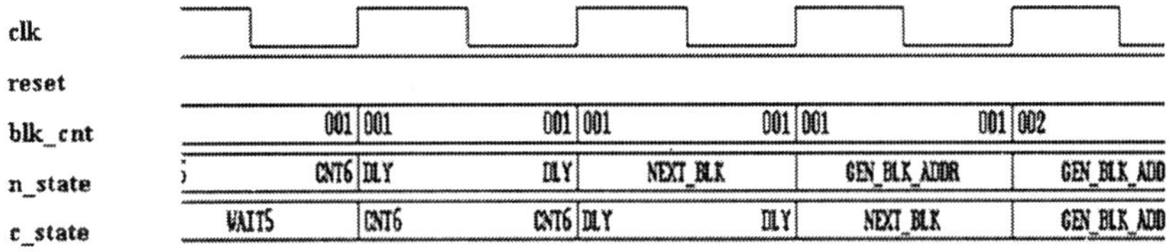

Figure 3-3. Waveform B for FSM1

Figure 3-3 shows that the FSM will loop back to GEN_BLK_ADDR state from the NEXT_BLK state if the register "blk_cnt" has not reached the value of 4096. When "blk_cnt" reaches the value of 4096, the FSM will break the loop from NEXT_BLK state and go to IDLE state.

3.1.2 SVA Checkers for FSM1

To verify FSM1 thoroughly, the following checks need to be done.

FSM1_chk1: FSM1 will always stay one-hot irrespective of the input conditions.

FSM1 is based on one-hot coding and hence should always have only one state bit asserted. If not, the FSM is not truly one-hot and the control signals might not be generated as expected. This can be tested by using any one of the built-in tasks, namely, **$countones** or **$onehot** defined in the SVA language.

```
property p_onehot;
  @(posedge clk) (reset_) |->
         ($countones(n_state) == 1);
endproperty

a_onehot: assert property(p_onehot);
c_onehot: cover property(p_onehot);
```

FSM1_chk2: If the current state is "IDLE" and if "get_data" is asserted, then the next state is "GEN_BLK_ADDR," and 64 cycles later the next state should be "WAIT0."

The FSM starts the transition, based on the IDLE state and the "get_data" signal. Once the FSM reaches GEN_BLK_ADDR, it has to stay there for 64 clock cycles.

```
sequence s_trans1;
     (c_state == IDLE) ##1
     ((c_state   ==    GEN_BLK_ADDR)    [*64])    ##1
     (c_state == WAIT0);
endsequence

property p_trans;
  @(posedge clk)
 (reset_ && $rose(get_data)) |->
           (reset_) throughout (s_trans1);
endproperty

a_trans: assert property (p_trans);
c_trans: assert property (p_trans)
```

The sequence "s_trans1" verifies that, if the current state of FSM1 is IDLE, then one cycle later it will transition to the GEN_BLK_ADDR. The FSM will stay in the state GEN_BLK_ADDR for 64 cycles (verified by using the repeat (*) operator) and one cycle later will move to the WAIT0 state. It is required that the reset is inactive throughout this property.

Figure 3-4 shows the results of "a_trans" property. The checker becomes active when there is a rising edge on the "get_data" signal and a match on the success is shown at the same point in the waveform. Though the checker stays active until reaching the WAIT0 state, the success is shown only at the starting point of the checker. The checker looks for a rising edge of "get_data" signal on every positive edge of the clock. If there isn't one, then the checker is assumed to succeed by default. This is a ***vacuous success*** as discussed in Chapter 1.

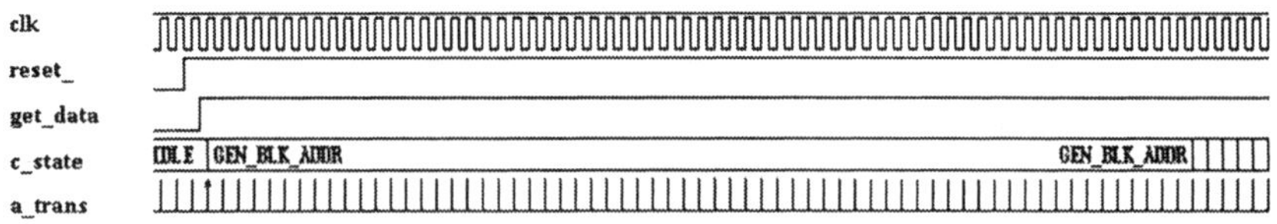

Figure 3-4. Waveform for FSM1_chk2

FSM1_chk3: If the current state is "WAIT0," then the FSM will transition states in a sequential manner from one state to another after one clock cycle each, unless the FSM is reset.

There is a wait state between every CNT* state. Hence, the FSM takes 2 clock cycles to move from one CNT* state to another CNT* state. The FSM moves in a linear fashion from CNT1 state to CNT6 state. The only possible path is as follows:

CNT1 -> CNT2 -> CNT3 -> CNT4 -> CNT5 -> CNT6

```
sequence s_trans3;
  ##1 (c_state == CNT1) ##2 (c_state == CNT2)
  ##2 (c_state == CNT3) ##2 (c_state == CNT4) ##2
  (c_state == CNT5) ##2 (c_state == CNT6);
endsequence

property p_linear_trans;
@(posedge clk)
((reset_) && (c_state == WAIT0)
&& ($past(c_state)== GEN_BLK_ADDR)) |->
```

```
s_trans3;
endproperty

a_linear_trans: assert property (p_linear_trans);
c_linear_trans: cover property (p_linear_trans);
```

Sequence “s_trans3” verifies that, if the FSM is currently in WAIT0 state and if it was in GEN_BLK_ADDR state in the previous cycle, then the FSM moves to CNT1 state after one cycle and WAIT1 state one cycle after that and so on up to reaching the state CNT6. Figure 3-5 shows the simulation results of the property p_linear_trans.

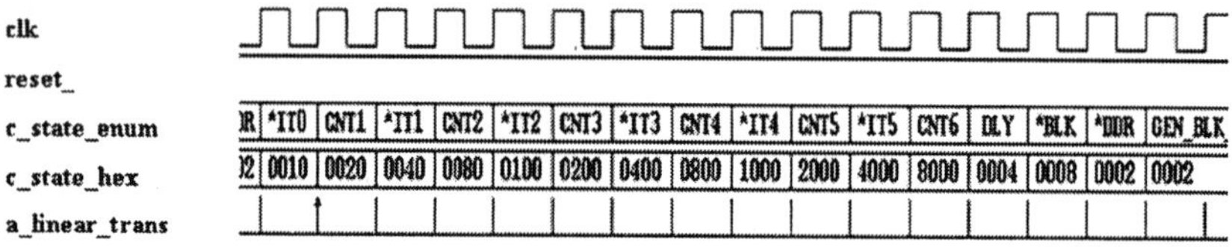

Figure 3-5. Waveform for FSM1_chk3

FSM1_chk4: Make sure that FSM1 is exercised such that it goes to both IDLE and GEN_BLK_ADDR at least once from the NEXT_BLK state.

This check acts as a functional coverage piece making sure that all paths of the FSM transitions are exercised once by the input test vectors.

```
sequence s_trans2;
  ##63 (c_state == GEN_BLK_ADDR) ##1
  (c_state == WAIT0);
endsequence

property p_frame;
  @(posedge clk)
 ((reset_) && (c_state == GEN_BLK_ADDR) &&
 (($past (c_state)== IDLE) ||
 ($past(c_state == NEXT_BLK)))) |->
                    s_trans2 ##0 s_trans3;
endproperty

a_frame: assert property(p_frame) cnt++;
```

```
c_frame: cover property(p_frame);

property p_complete_frame;
 @(posedge clk)
((cnt == 16'd4095)&&reset_&&
(c_state==CNT6)) |->
                        done_frame;
endproperty

a_complete_frame:
   assert property(p_complete_frame)
   $display ("A complete frame has been
   transferred \n");
```

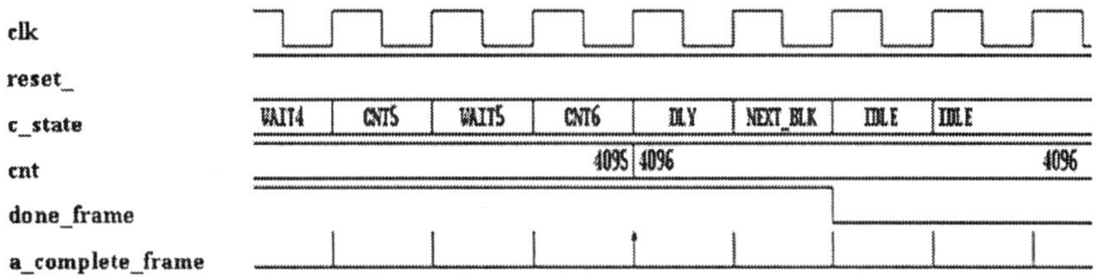

Figure 3-6. Waveform for FSM1_chk4

The property "p_frame" verifies the complete state transition of FSM1 starting from IDLE state. While this check is performed, a local variable "cnt" is incremented in the action block every time the property "a_frame" succeeds. When the value of the variable "cnt" reaches 4095, all blocks of data have been processed and the control signal "done_frame" is asserted. At the end of a complete frame testing, the action block of the check "a_complete_frame" can be used to display the results. Figure 3-6 shows the simulation results of the property p_complete_frame.

Two separate properties "p_frame_path1" and "p_frame_path2" are written to make sure that all possible paths of the FSM are covered during simulation.

```
property p_frame_path1;
  @(posedge clk)
  ((reset_) && (c_state == GEN_BLK_ADDR) &&
  ($past(c_state == NEXT_BLK)) |->
  s_trans2 ##0 s_trans3;
```

```
endproperty

c_frame_path1: cover property (p_frame_path1);

property p_frame_path2;
  @(posedge clk)
  ((reset_) && (c_state == GEN_BLK_ADDR) &&
  ($past(c_state == IDLE))) |->
     s_trans2 ##0 s_trans3;
endproperty

c_frame_path2: cover property (p_frame_path2);
```

3.2 Sample Design – FSM2

A slightly more complicated FSM is discussed in this section. The FSM discussed in Section 3.1 was linear and did not have many ways of getting to a particular state. FSM2 will have fewer states but there will be more ways of getting to a particular state. This presents a minor challenge in extracting the checks that need to be done.

3.2.1 Functional description of FSM2

FSM2 performs the role of an arbiter. At any given time, FSM2 can arbitrate between 3 master devices. Any or all of the master devices can request for the grant of the bus and the arbiter will decide who gets the bus based on a round robin fashion. Once the master acquires a grant, it uses the bus to do certain transactions. At the end of the transaction, the master lets the arbiter know and the bus is freed. Once the bus is free, all the masters can once again make a request for the bus if they have any pending transactions. The key concept is to make sure that the arbiter is not starving any of the masters.

The FSM has 7 possible states shown as follows:

IDLE = 7'b0000001
MASTER1 = 7'b0000010
IDLE1 = 7'b0000100
MASTER2 = 7'b0001000
IDLE2 = 7'b0010000
MASTER3 = 7'b0100000
IDLE3 = 7'b1000000

The FSM is coded with a one-hot coding style. Figure 3-7 shows the bubble diagram of FSM2.

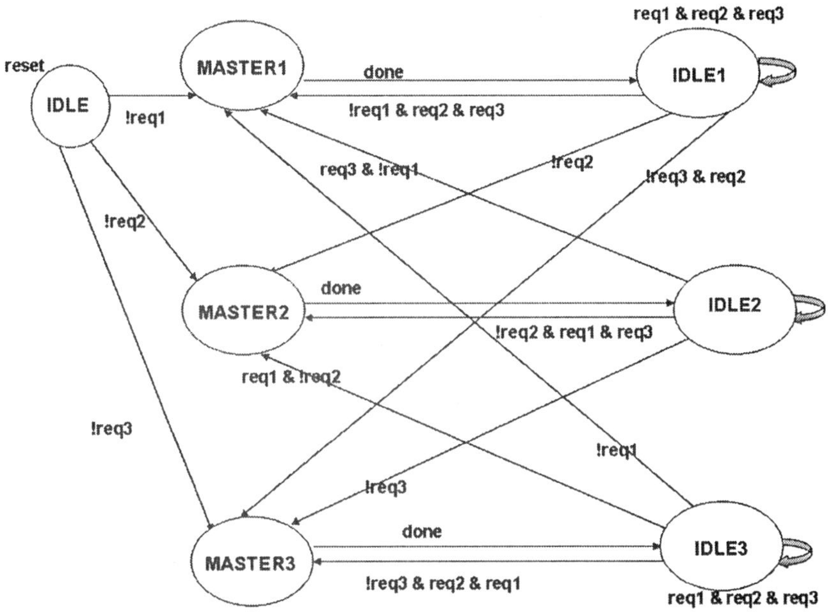

Figure 3-7. Bubble diagram for FSM2

- Upon reset the state machine moves to the IDLE state.
- While the state machine is in the IDLE state, it looks for a "req" from any of the master devices wanting to use the bus. The grant is given in a priority-encoded fashion. For example, when the FSM is in IDLE state, if all three master devices make a request then "master1" gets the bus.
- While the state machine is in the MASTER* state, it asserts the "gnt" signal of the respective master device.
- Once the master device is done with using the bus, it indicates this to the arbiter by asserting the "done" signal and this moves the FSM to the respective IDLE* state of that master device.
- When the FSM is in the IDLE1 state, it will look for "req" from the masters in the order of master2, master3 and then master1.
- When the FSM is in the IDLE2 state, it will look for "req" from the masters in the order of master3, master1 and then master2.

- When the FSM is in the IDLE3 state, it will look for "req" from the masters in the order of master1, mastcr2 and then master3.

Example 3.2 FSM2 Sample code

```
module  bus_arbiter(clk,  reset,  frame,  irdy,
req1, req2, req3, gnt1, gnt2, gnt3);

input logic clk, reset, frame, irdy;
input logic req1, req2, req3;

output gnt1, gnt2, gnt3;

enum bit [6:0] {IDLE = 7'b0000001,
   MASTER1 = 7'b0000010,
   IDLE1 = 7'b0000100,
   MASTER2 = 7'b0001000,
   IDLE2 = 7'b0010000,
   MASTER3 = 7'b0100000,
   IDLE3 = 7'b1000000} next, state;

logic done, gnt1, gnt2, gnt3;

/* define glue signals */

assign done = frame && irdy;

/* state register code */

always@(posedge clk or negedge reset)
begin
  if(!reset)
    state <= IDLE;
  else
    state <= next;
end

/* next state combinational logic */

always@(*)
begin
  next = IDLE;
  case(state)
```

```
IDLE:
  if (req1 == 1'b0)
    next <= MASTER1;
  else if (req1 == 1'b1 & req2 == 1'b0)
    next <= MASTER2;
  else if (req3 == 1'b0 & req1 == 1'b1)
    next <= MASTER3;
  else
    next <= IDLE;

MASTER1:
  if(!done)
    next <= MASTER1;
  else
    next <= IDLE1;

IDLE1:

  if(req2 == 1'b0 )
    next <= MASTER2;
  else if (req3 == 1'b0 & req2 == 1'b1)
    next <= MASTER3;
  else if (req3 == 1'b1 & req1 == 1'b0 & req2 ==
  1'b1)
    next <= MASTER1;
  else
    next <= IDLE1;

MASTER2:
  if(!done)
    next <= MASTER2;
  else
    next <= IDLE2;

IDLE2:
  if (req3 == 1'b0)
    next <= MASTER3;
  else if (req3 == 1'b1 & req1 == 1'b0)
    next <= MASTER1;
  else if (req1 == 1'b1 & req2 == 1'b0)
    next <= MASTER2;
  else
```

```
      next <= IDLE2;
  MASTER3:
    if (!done)
      next <= MASTER3;
    else
      next <= IDLE3;

  IDLE3:
    if (req1 == 1'b0)
      next <= MASTER1;
    else if (req1 == 1'b1 & req2 == 1'b0)
      next <= MASTER2;
    else if (req2 == 1'b1 & req3 == 1'b0)
      next <= MASTER3;
    else
      next <= IDLE3;
  endcase

end

/* output generating statements */

assign gnt1 = ((state == MASTER1)) ? 0 : 1;
assign gnt2 = ((state == MASTER2)) ? 0 : 1;
assign gnt3 = ((state == MASTER3)) ? 0 : 1;

endmodule
```

Figure 3-8 shows a sample waveform for FSM2. For convenience, the state encoding is shown both in the enumerated value and hexadecimal value. The state value *1 means that the state is MASTER1, similarly, *2 for MASTER2 and *3 for MASTER3. At marker 1, both master2 and master3 make a request for the bus. The FSM is in IDLE3 state at this point and hence provides the grant to master2. At marker 2, the FSM is in IDLE2 state and both master1 and master3 request the bus. This time master3 gets the grant. The grant provided always depends on which IDLE* state the FSM is currently in.

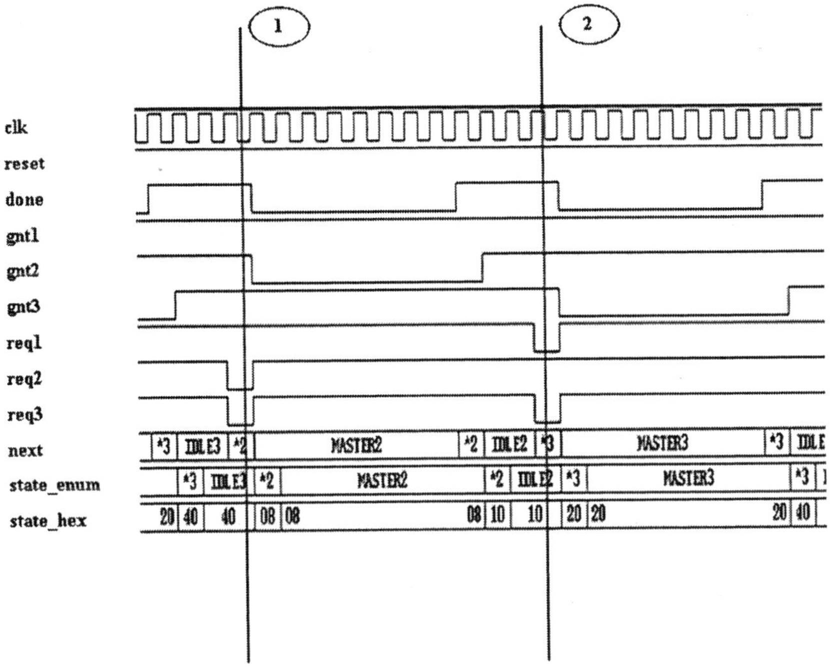

Figure 3-8. Waveform for FSM2

3.2.2 SVA Checkers for FSM2

For a state machine like this, since there are many ways to get to a specific state, the first thing to understand is what are the possible legal paths? This can be a very difficult process depending on the complexity of the state machine. As a first step, a matrix should be created with all states represented on both the x and y axis. Then the matrix should be filled with a "Yes" or "No" indicating whether it is possible to transition from that state in the x axis to a respective state in the y axis. Once we have a representation as described above, we can start categorizing the SVA checks.

- Based on the matrix, if a state to state transition is forbidden, then it should be verified using a SVA check.
- All possible legal state transitions should be covered by the testbench. This metric can be measured by using the "cover" statements on SVA properties. The same information can also be obtained from code coverage tools.

Based on the matrix analysis, as shown in Table 3-1, the following checks need to be written to verify FSM2 thoroughly.

Table 3-1. Matrix diagram for FSM2 state transition

	IDLE	M1	I1	M2	I2	M3	I3
IDLE	Y	Y	N	Y	N	Y	N
M1	Y	Y	Y	N	N	N	N
I1	Y	Y	Y	Y	N	Y	N
M2	Y	N	N	Y	Y	N	N
I2	Y	Y	N	Y	Y	Y	N
M3	Y	N	N	N	N	Y	Y
I3	Y	Y	N	Y	N	Y	Y

FSM2_chk1: FSM2 should always behave as a one-hot state machine.

```
property p_fsm2_encoding;
  @(posedge clk) $onehot(state);
endproperty

a_fsm2_encoding:
          assert property (p_fsm2_encoding);
c_fsm2_encoding:
          cover property (p_fsm2_encoding);
```

The built-in function **$onehot** can be used to make sure that only one bit of the state register is high at any given time, thus proving that the FSM always stays one-hot.

FSM2_chk2: From IDLE state the FSM cannot go to IDLE1, IDLE2 or IDLE3 states.

```
property p_forbid_trans1;
@(posedge clk)
(((state == IDLE1) || (state == IDLE2) ||
(state == IDLE3)) && reset) |->
        $past ((state == IDLE) == 0);
endproperty

a_forbid_trans1:assert property(p_forbid_trans1);
```

The property "p_forbid_trans1" verifies that, if the current state is IDLE1, IDLE2 or IDLE3, then the state of the FSM in the previous cycle cannot be IDLE.

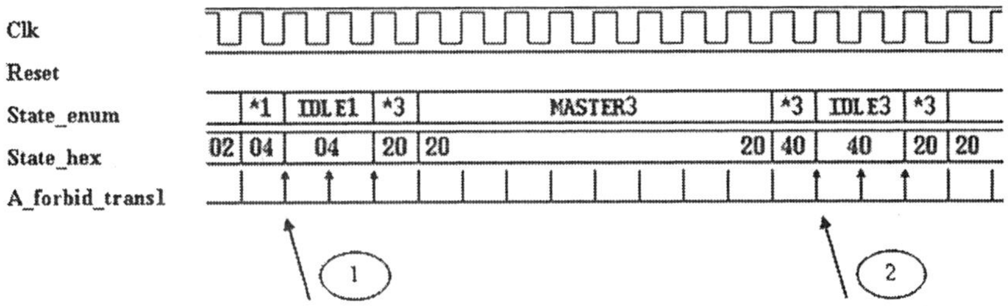

Figure 3-9. Waveform for FSM2_chk2

Figure 3-9 shows the results of the check "a_forbid_trans1." Marker 1 shows a point where the FSM is currently in IDLE1 state. The property passes here since the FSM was in MASTER1 state in the previous cycle and not in IDLE state. Similarly, marker 2 shows a point where the FSM is in IDLE3 state. The property passes here also since the FSM was in MASTER3 state in the previous clock cycle and not IDLE state.

FSM2_chk3:

From MASTER1 state the FSM cannot go to other MASTER states or IDLE2 or IDLE3.

From MASTER2 state the FSM cannot go to other MASTER states or IDLE1 or IDLE2.

From MASTER3 state the FSM cannot go to other MASTER states or IDLE1 or IDLE2.

This check makes sure that, if the FSM is in a certain MASTER* state, then the next transition will always be to the IDLE* state specific to the master state. In other words, if the FSM is currently in MASTER1 state, then it has to transition to only IDLE1 state next assuming the FSM is not reset. If it transitions to any other state, it is a violation. Similarly, MASTER2 should transition to IDLE2 and MASTER3 to IDLE3 state respectively. Figure 3-10 shows the result of the check "a_forbid_trans2a."

```
property p_forbid_trans2a;
  @(posedge clk)
  (((state == IDLE2) || (state == IDLE3) ||
  (state == MASTER2) || (state == MASTER3))
  && reset) |->
              $past ((state == MASTER1) == 0);
endproperty

a_forbid_trans2a:
         assert property(p_forbid_trans2a);
c_forbid_trans2a:
         cover property(p_forbid_trans2a);

property p_forbid_trans2b;
  @(posedge clk)
  (((state == IDLE1) || (state == IDLE3) ||
  (state == MASTER1) || (state == MASTER3))
  && reset) |->
              $past ((state == MASTER2) == 0);
endproperty

a_forbid_trans2b:
          assert property(p_forbid_trans2b);
c_forbid_trans2b:
          cover property(p_forbid_trans2b);

property p_forbid_trans2c;
  @(posedge clk)
  (((state == IDLE2) || (state == IDLE1) ||
  (state == MASTER2) || (state == MASTER1))
  && reset) |->
         $past (state == MASTER3) == 0);
endproperty

a_forbid_trans2c:
         assert property(p_forbid_trans2c);
c_forbid_trans2c:
         cover property(p_forbid_trans2c);
```

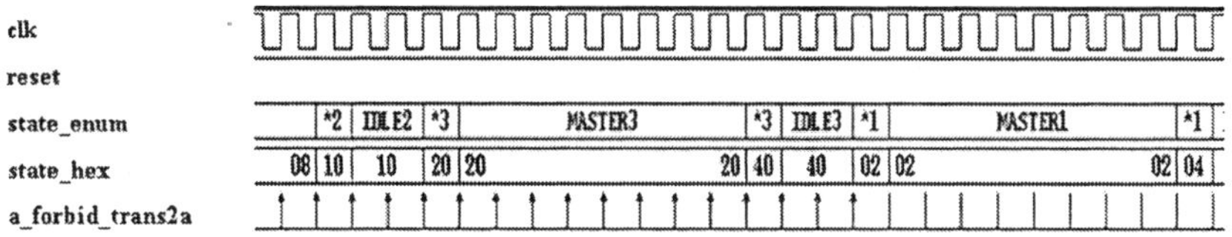

Figure 3-10. Waveform for FSM2_chk3

FSM2_chk4:
From IDLE1 state the FSM cannot go to IDLE2 or IDLE3 states.
From IDLE2 state the FSM cannot go to IDLE3 or IDLE1 states.
From IDLE3 state the FSM cannot go to IDLE2 or IDLE1 states.

This check makes sure that the FSM will always transition to a MASTER* state from an IDLE* state assuming that the FSM is not reset in between. If the FSM transitions from one IDLE* state to another IDLE* state, it is a violation. Figure 3-11 shows the result of the check "p_forbid_trans3a."

```
property p_forbid_trans3a;
  @(posedge clk)
  (((state == IDLE2) || (state == IDLE3))
  && reset) |->
        $past (state== IDLE1) == 0);
endproperty

a_forbid_trans3a:
           assert property(p_forbid_trans3a);
c_forbid_trans3a:
           cover property(p_forbid_trans3a);

property p_forbid_trans3b;
  @(posedge clk)
  (((state == IDLE1) || (state == IDLE3))
  && reset) |->
         $past (state== IDLE2) == 0);
endproperty

a_forbid_trans3b:
           assert property(p_forbid_trans3b);
c_forbid_trans3b:
           cover property(p_forbid_trans3b);
```

```
property p_forbid_trans3c;
  @(posedge clk)
  (((state == IDLE1) || (state == IDLE2))
  && reset) |->
            $past (state== IDLE3) == 0);
endproperty

a_forbid_trans3c:
            assert property(p_forbid_trans3c);
c_forbid_trans3c:
            cover property(p_forbid_trans3c);
```

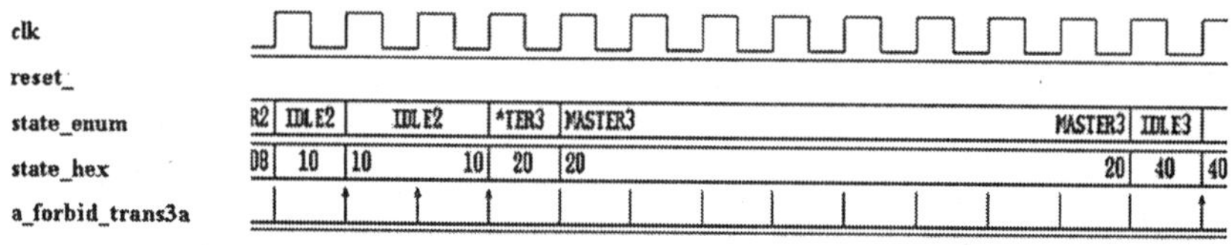

Figure 3-11. Waveform for FSM2_chk4

FSM2_chk5: There should be a grant for every request sent to the arbiter.

```
property p_req_gnt;
  @(posedge clk)
 ((!req1 || !req2 || !req3) && reset) |->
           ##1 (!gnt1 || !gnt2 || !gnt3);
endproperty

a_req_gnt: assert property(p_req_gnt);
c_req_gnt: cover property(p_req_gnt);
```

The property "p_req_gnt" verifies that, if any of the masters make a request for the bus, then within one cycle, any one of the "gnt" signal should be asserted. If the grant does not arrive in one cycle, it is a fatal error.

Figure 3-12 shows the result of the check 'a_req_gnt." Marker 1 shows the point in the waveform where master3 is requesting the bus and within one cycle the signal "gnt3" is asserted and hence the check passes. Similarly, marker 2 is pointing to a place where both master2 and master3 are requesting the bus and "gnt2" is asserted within one clock cycle and hence the check passes.

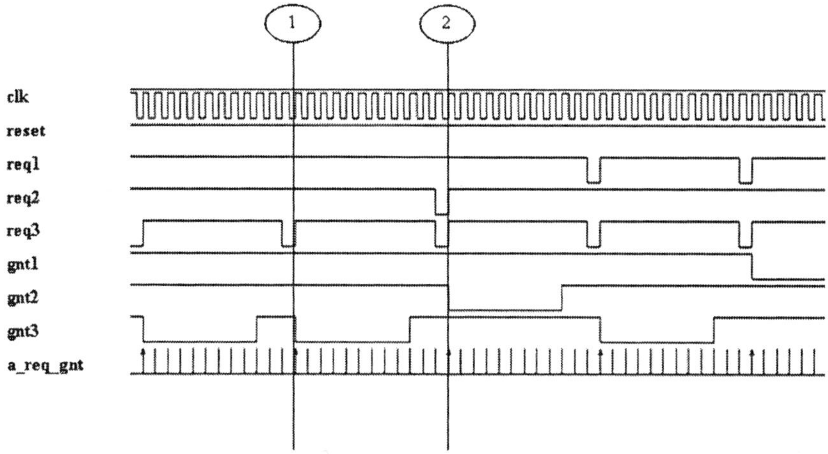

Figure 3-12. Waveform for FSM2_chk5

FSM2_chk6: Check for the fairness of the arbiter. Make sure that all the masters are getting equal number of grants.

```
property p_req_gnt_1;
  @(posedge clk) ((!req1 && reset)) |->
          ##1 !gnt1;
endproperty

c_req_gnt_1: cover property(p_req_gnt_1);

property p_req_gnt_2;
  @(posedge clk) ((!req2 && reset)) |->
          ##1 !gnt2;
endproperty

c_req_gnt_2: cover property(p_req_gnt_2);

property p_req_gnt_3;
  @(posedge clk) ((!req3 && reset)) |->
          ##1 !gnt3;
endproperty
```

```
c_req_gnt_3: cover property(p_req_gnt_3);

property p_req1;
 @(posedge clk)  ($fell(req1) && reset);
endproperty

c_req1: cover property(p_req1);

property p_req2;
 @(posedge clk)  ($fell(req2) && reset);
endproperty

c_req2: cover property(p_req2);

property p_req3;
 @(posedge clk)  ($fell(req3) && reset);
endproperty

c_req3: cover property(p_req3);
```

This check is performed to get the functional coverage information and also to validate the fairness of the arbiter. Three properties p_req_gnt1, p_req_gnt2 and p_req_gnt_3 are written to calculate how many times a master was actually able to get a grant. The next three properties, p_req1, p_req2 and p_req3 are written to calculate how many requests each master actually made. By using the cover statements on these properties, the simulation results are printed based on the number of matches. In a sample random test environment, the following results were produced

```
c_req_gnt_1, 10433 attempts, 288 match
c_req_gnt_2, 10433 attempts, 290 match
c_req_gnt_3, 10433 attempts, 291 match
c_req1, 10433 attempts, 481 match
c_req2, 10433 attempts, 474 match
c_req3, 10433 attempts, 505 match
```

Note that, each master requested the bus approximately 475 times and each one of the masters was granted the bus approximately 290 times. This shows that the arbiter is being very fair and is not starving any one master device.

3.2.3 FSM2 with a timing window protocol

In the previous section, FSM2 asserted the "gnt" signal one clock cycle after a request was made. In this section, the arbiter functionality is assumed such that it can take anywhere between 2 to 5 clock cycles to produce a grant. While most of the protocol extraction process still remains the same for the new arbiter, the timing needs to be adjusted in some of the checks.

```
assign req = !req1 || !req2 || !req3;
assign gnt = !gnt1 || !gnt2 || !gnt3;

property p_req_gnt_w;
    @(posedge clk) $rose(req) |->
                    ##[2:5] $rose(gnt);
endproperty

a_req_gnt_w : assert property(p_req_gnt_w);
```

The property p_req_gnt_w looks for a rising edge on the "req" signal. The "req" signal is the OR output of all the three requests req1, req2 and req3 respectively. Once the pre-condition is true, it verifies that a rising edge occurs on the "gnt" signal within 2 to 5 clock cycles. The "gnt" signal is the OR output of all the three "gnt" signals gnt1, gnt2 and gnt3 respectively. Functional coverage statements similar to the ones shown in check FSM2_chk6 can be easily written for the new protocol based on the window of time. Figure 3-13 shows the results of the check "a_req_gnt_w."

The marker "1s" indicates the first valid request made to the arbiter. A valid "gnt" comes at marker "1e" after 5 clock cycles and hence the checker passes. The marker "2s" indicates the second valid request made to the arbiter. A valid "gnt" comes at the marker "2e" after 2 clock cycles and hence the checker passes.

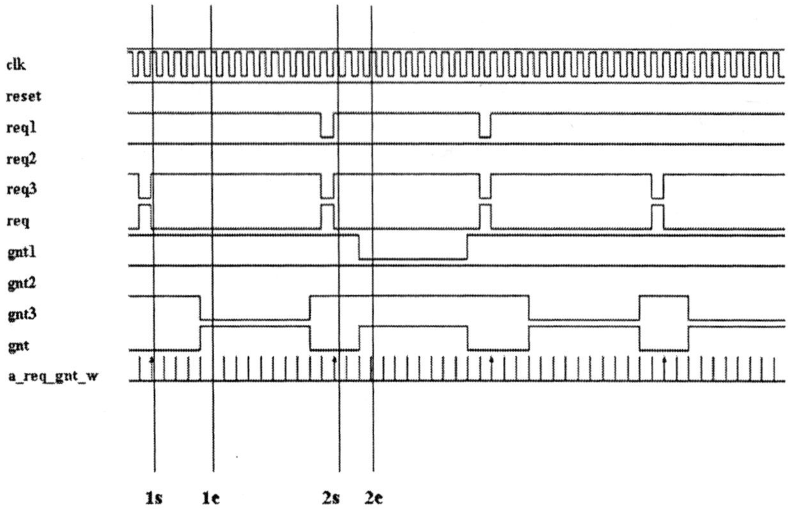

Figure 3-13. Waveform for window check

A key functional coverage data that a user might be interested in is the arbiter latency. The arbiter can take 2 to 5 cycles to respond to each one of the master devices. It is important to know if the average latency of the arbiter is the same for all three masters. We can use SVA cover statements to calculate the response time of the arbiter for each master.

```
genvar s;
generate

for (s=2; s<6; s++)
begin: generic
c_gnt_generic :
cover property(@(posedge clk) $rose(gnt) |->
          ($past(req,s) == 1'b1));
end
endgenerate
```

A generate statement can be used to create an array of cover statements. The objective is to find out the average response time of the arbiter. The built-in function **$past** is used to define a valid request and grant sequence. The variable "s" is used to loop around and create 4 separate cover properties for each possible latency values (2, 3, 4 and 5 clock cycles respectively). The cover statement will increment the appropriate latency bin. Sample simulation results produced are shown below.

```
  tb.u4.u1.generic[2].c_gnt_generic, 5793 attempts,
112 match, 0 vacuous match
  tb.u4.u1.generic[3].c_gnt_generic, 5793 attempts,
113 match, 0 vacuous match
  tb.u4.u1.generic[4].c_gnt_generic, 5793 attempts,
101 match, 0 vacuous match
  tb.u4.u1.generic[5].c_gnt_generic, 5793 attempts,
104 match, 0 vacuous match
```

From these results, it is clear that the arbiter has an even distribution of latency. The previous example can be slightly modified to get latency information specific to each master. This way we will know if it takes longer to provide a grant to any specific master.

```
assign req_local[3:1] = ({req3, req2, req1});
assign gnt_local[3:1] = ({gnt3, gnt2, gnt1});

genvar j, k;
generate

for (j=2; j<6; j++)
begin: latency
for (k=1; k<4; k++)
begin: Master
c_gnt_o :
cover property(@(posedge clk) $fell(gnt_local[k])
          |-> ($past(req_local[k],j) == 1'b0));
end
end
endgenerate
```

Note that, a vector of the "gnt" signals called "gnt_local" and a vector of the "req" signals called "req_local" are defined. This allows one to loop through each master one at a time. Two loops are used, the outer loop "latency" defines the latency bins and the inner loop "Master" defines the master's identity. The property is active once a valid "gnt" signal is detected. A valid "req" for this specific 'gnt" is searched in the past anywhere between 2 and 5 clock cycles. Sample simulation results for such a coverage statement is shown below.

```
  tb.u4.u1.latency[2].Master[1].c_gnt_o,           5793
attempts, 41 match, 0 vacuous match
  tb.u4.u1.latency[2].Master[2].c_gnt_o,           5793
attempts, 37 match, 0 vacuous match
  tb.u4.u1.latency[2].Master[3].c_gnt_o,           5793
attempts, 34 match, 0 vacuous match
  tb.u4.u1.latency[3].Master[1].c_gnt_o,           5793
attempts, 39 match, 0 vacuous match
  tb.u4.u1.latency[3].Master[2].c_gnt_o,           5793
attempts, 34 match, 0 vacuous match
  tb.u4.u1.latency[3].Master[3].c_gnt_o,           5793
attempts, 40 match, 0 vacuous match
  tb.u4.u1.latency[4].Master[1].c_gnt_o,           5793
attempts, 27 match, 0 vacuous match
  tb.u4.u1.latency[4].Master[2].c_gnt_o,           5793
attempts, 36 match, 0 vacuous match
  tb.u4.u1.latency[4].Master[3].c_gnt_o,           5793
attempts, 38 match, 0 vacuous match
  tb.u4.u1.latency[5].Master[1].c_gnt_o,           5793
attempts, 34 match, 0 vacuous match
  tb.u4.u1.latency[5].Master[2].c_gnt_o,           5793
attempts, 29 match, 0 vacuous match
  tb.u4.u1.latency[5].Master[3].c_gnt_o,           5793
attempts, 41 match, 0 vacuous match
```

3.3 Summary on SVA for FSM

- FSMs are an integral part of any design and they need to be verified thoroughly.
- Every forbidden transition should be checked using SVA. If a forbidden transition occurs, it should be flagged as a fatal error.
- The testbench must cover all possible legal transitions. Functional coverage information should be used wisely to build a reactive simulation environment.

Chapter 4

SVA FOR DATA INTENSIVE DESIGNS

In any design, there are two areas that need to be verified thoroughly:

a. Is the control logic behaving correctly? – These signals control the flow of data in the design and have complex timing relationships between each other.
b. Is my output data as expected? – This makes sure that the output data of the RTL matches the output of the golden model (usually written in C). This guarantees that the functionality of the optimized hardware algorithms implemented in RTL matches that of the golden model.

In general, assertion based verification is very suited for checking signals that have complex timing relationships or in other words, the control logic. The declarative nature of the language makes it more suitable for temporal checking. While assertions don't add any additional value for data checking, it can still be used for writing efficient self-checking environments.

4.1 A simple multiplier check

SystemVerilog assertions have the advantage of using most data types and operators that are part of the SystemVerilog language. This gives great flexibility in writing simple arithmetic checks.

Example 4.1 A simple multiplier

```
module au (
  input logic [7:0] a, b, c,
```

```
  input logic sel,
  output logic [15:0] d
);

logic [15:0] e;
logic [15:0] sel_h, sel_l;

// Resource sharing architecture

always_comb
begin
  if(sel) e = b; else e = c;
  d = multiply(a, e);
end

// Functional sva checker

always@(a, b, c, sel)
begin
sel_h = a*b;
sel_l = a*c;

if(sel)
sel_high : assert (sel_h == d);
if (!sel)
sel_low : assert (sel_l == d);

end
endmodule
```

Example 4.1 shows a simple multiplier. Only one multiplier is used and the multiplication is done based on the input that is selected by the "sel" line. Two simple checkers named "sel_high" and "sel_low" can be written to verify the multiplier. The check "sel_high" is active when the "sel" line is high and the check "sel_low" is active when the "sel" line is low. The user can choose to use any type of multiplier relevant to the user's environment. For example, it can be a shift/add multiplier, a booth multiplier or something else. From a functional verification standpoint, we need to make sure that no matter which type of multiplier algorithm is used, the end output result matches. Figure 4-1 shows the results produced by these two checkers. Note that the checker is active, based on the status of the "sel" signal (immediate assertion).

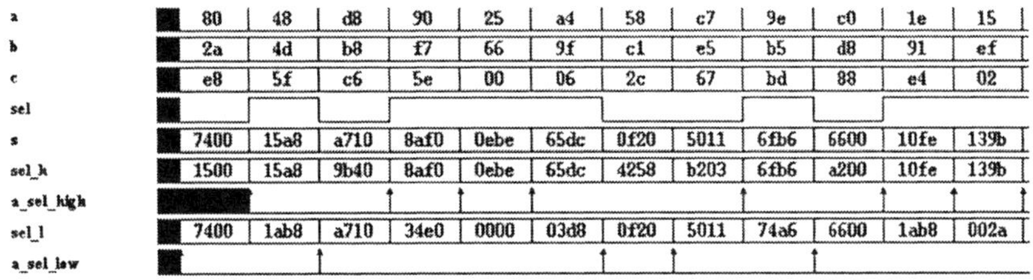

Figure 4-1. Waveform for Multiplier checker

4.2 Sample Design – Arithmetic unit

In this section, verification of an arithmetic unit is discussed (a Walsh-Hadamard Transform (WHT) block). WHT is a common algorithm used in still image compression applications. WHT is used to convert the pixels from time domain to frequency domain before the image is encoded. Typically, these algorithms are tested very easily with C or Matlab programs. But when the algorithms are converted to hardware, it goes through severe optimizations that will make it more hardware efficient. It is very common to provide the same input data to both the golden C model and the RTL model. The RTL is verified by comparing its output with that of the C model. In this section, SVA is used to produce the golden results dynamically during a simulation and compared with the results from RTL.

4.2.1 WHT Algorithm

The WHT algorithm is an 8x8 matrix multiplication. In image compression applications, data is processed one block at a time. Each block is an 8x8 matrix, or in other words, 64 data points. The objective is to perform a matrix multiplication between 2 matrices, each of size 8x8, to produce the end result, an 8x8 matrix. Due to the repetitive nature of the matrix multiplication, this can be achieved one block at a time. The WH matrix is defined as follows. Since the matrix without the scaling factor consists of +1 and –1, the transform operation consists simply of addition and subtraction.

```
WHT [8][8] =
    {
        {1, 1, 1, 1, 1, 1, 1, 1},
        {1, 1, 1, 1,-1,-1,-1,-1},
        {1, 1,-1,-1,-1,-1, 1, 1},
        {1, 1,-1,-1, 1, 1,-1,-1},
```

```
{1,-1,-1, 1, 1,-1,-1, 1},
{1,-1,-1, 1,-1, 1, 1,-1},
{1,-1, 1,-1,-1, 1,-1, 1},
{1,-1, 1,-1, 1,-1, 1,-1}

};
```

4.2.2 WHT Hardware implementation

In hardware, the WHT algorithm can be optimized to reduce the number of additions and subtractions performed. This is achieved by exploiting the redundant additions and subtractions performed on the same set of data. Each arithmetic block is optimized to perform a 1x8 by 8x8 matrix multiplication. In other words, one row of data (8 data points) is processed at a time. The simplified arithmetic unit performs 3 stages of addition and subtraction to produce one row of output data. Figure 4-2 shows the block diagram of the hardware implementation of the WHT algorithm.

Assume D1, D2, D3, D4, D5, D6, D7 and D8 form a row of data. Now this row of data has to be processed through the WHT matrix. The equations of the three stages of optimized arithmetic operations can be listed as follows.

Stage 1

$Y1 = D1 + D2$, $Y2 = D3 + D4$, $Y3 = D5 + D6$, $Y4 = D7 + D8$,
$Y5 = D1 + D4$, $Y6 = D5 + D8$, $Y7 = D2 + D3$, $Y8 = D6 + D7$,
$Y9 = D1 + D3$, $Y10 = D6 + D8$, $Y11 = D2 + D4$, $Y12 = D5 + D7$

Stage 2

$Z1 = Y1 + Y2$, $Z2 = Y3 + Y4$, $Z3 = Y1 + Y4$, $Z4 = Y2 + Y3$,
$Z5 = Y1 + Y3$, $Z6 = Y2 + Y4$, $Z7 = Y5 + Y6$, $Z8 = Y7 + Y8$,
$Z9 = Y5 + Y8$, $Z10 = Y7 + Y6$, $Z11 = Y9 + Y10$,
$Z12 = Y11 + Y12$, $Z13 = Y9 + Y12$, $Z14 = Y11 + Y10$

Stage 3

$X1 = Z1 + Z2$, $X2 = Z1 - Z2$, $X3 = Z3 - Z4$, $X4 = Z5 - Z6$,
$X5 = Z7 - Z8$, $X6 = Z9 - Z10$, $X7 = Z11 - Z12$, $X8 = Z13 - Z14$

X1, X2, X3, X4, X5, X6, X7 and X8 form a row of processed output data.

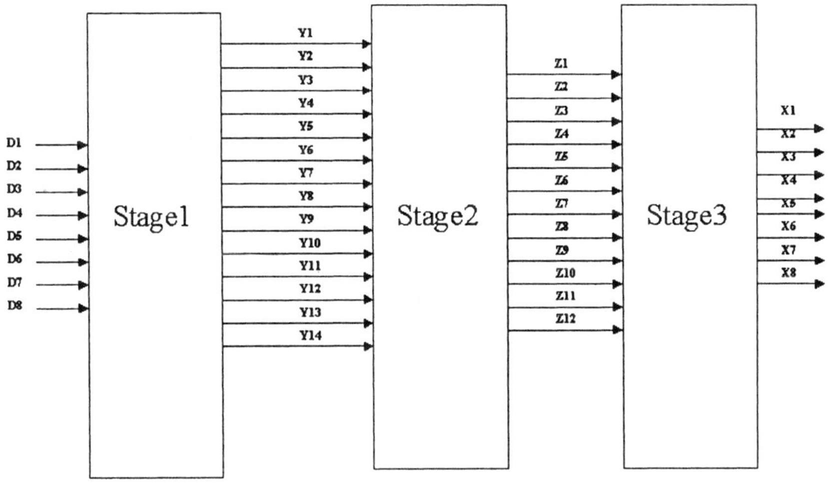

Figure 4-2. WHT hardware block diagram

4.2.3 SVA Checker for WHT block

To functionally verify this block, a checker does not have to know the internal implementation details of this block. A checker should be able to produce the golden result and compare it with the design data. The golden data can be produced within the SVA checker by simply performing a matrix multiplication. This data can then be compared with the output of the WHT block. Figure 4-3 shows a simple checker configuration for the WHT block.

It is very common to register the outputs of such combination blocks to obtain the most stable data. A checker can be written using the enable signal of the register as the trigger. The results produced within the checker should be compared with the original registered output of the WHT arithmetic block. A sample SVA checker is shown in Example 4.2.

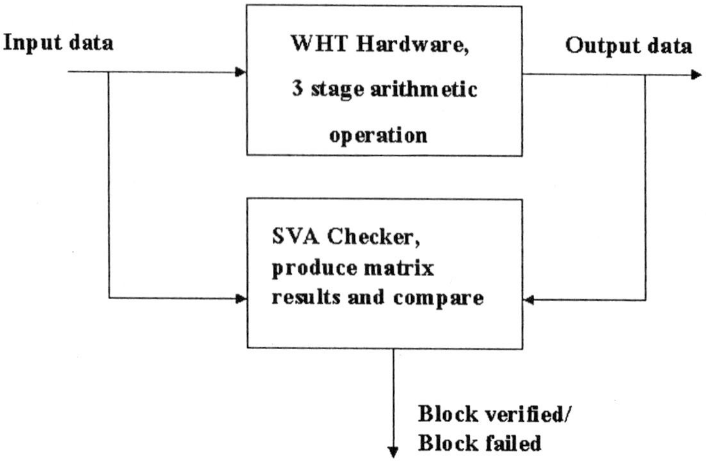

Figure 4-3. WHT checker configuration

Example 4.2 SVA Checker for WHT block

```
module au_comp_chk (
  input logic clk, reset, enable1, enable2,
  input logic signed [15:0]
        d1, d2, d3, d4, d5, d6, d7, d8,
  input logic signed [15:0]
        o1, o2, o3, o4, o5, o6, o7, o8
);

logic signed [15:0] in_local[0:7];
logic signed [15:0] out_orig[0:7];
logic signed [15:0] out_local[0:7];

integer i, k;

integer wh_local[0:7][0:7] =
    {
        {1, 1, 1, 1, 1, 1, 1, 1},
        {1, 1, 1, 1,-1,-1,-1,-1},
        {1, 1,-1,-1,-1,-1, 1, 1},
        {1, 1,-1,-1, 1, 1,-1,-1},
        {1,-1,-1, 1, 1,-1,-1, 1},
        {1,-1,-1, 1,-1, 1, 1,-1},
        {1,-1, 1,-1,-1, 1,-1, 1},
```

```
        {1,-1, 1,-1, 1,-1, 1,-1}

        };

always@(o1, o2, o3, o4, o5, o6, o8)
  begin
   out_orig[0] <= o1;
   out_orig[1] <= o2;
   out_orig[2] <= o3;
   out_orig[3] <= o4;
   out_orig[4] <= o5;
   out_orig[5] <= o6;
   out_orig[6] <= o7;
   out_orig[7] <= o8;
  end

  always@(d1, d2, d3, d4, d5, d6, d7, d8)
  begin

   for(i=0; i<8; i++)

   begin

   out_local[i] <=
  (d1*wh_local[i][0]) + (d2*wh_local[i][1]) +
  (d3*wh_local[i][2])    +    (d4*wh_local[i][3])    +
 (d5*wh_local[i][4]) +   (d6*wh_local[i][5])         +
 (d7*wh_local[i][6]) + (d8*wh_local[i][7]) ;

  end

  end

  genvar j;
  generate

  for(j=0; j<8; j++)
  begin : loop
  a_au_comp_chk_o :
  assert property
  (@(posedge clk) (reset &&enable2) |->
            (out_local[j] == out_orig[j]));
```

```
end

endgenerate

endmodule

bind au_comp au_comp_chk a1
(clk, reset, enable1, enable2, d1, d2, d3, d4,
d5, d6, d7, d8, o1, o2, o3, o4, o5, o6, o7,
o8);
```

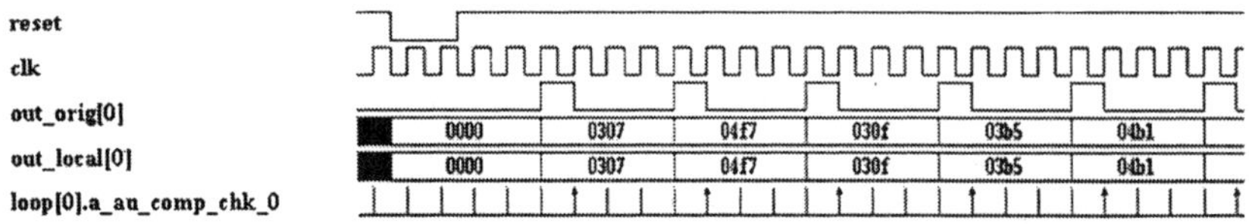

Figure 4-4. Waveform for WHT checker

The checker calculates the expected output locally and puts the results in an array named "out_local." The original output data from the design is also stored in an array named "out_orig." The checker creates an array of properties to verify the 8 data points by using the "generate" statement. The special variable "genvar," allows the use of a "for loop" to create 8 separate properties to verify each one of the data points simultaneously. The property is asserted when the enable signal is high and the design is not in reset. Each property will compare the respective output data, "out_local" and "out_orig." If they are not equal, the assertion fails. Figure 4-4 shows the results from the first data point in the array.

4.3 Sample Design – A JPEG based data-path design

In this section, verification of a sample JPEG model is discussed. The design block is part of a JPEG encoder wherein data is read from memory and transformed using certain arithmetic algorithms. The transformed data is then stored in a memory for package and transmission.

There are three main modules in the JPEG model - the data feeder, the data path and the data control modules. The top level block diagram of the JPEG model is shown in Figure 4-5. Details of each module are provided below:

- The data control module helps with the hand shaking process with the memories and also generates control signals for the data path and data feeder modules to process the data.
- The data feeder module reads a block of data at a time from the memory and provides it to the data path module to process.
- The data path module performs the arithmetic operation on the block of data and stores it in memory.

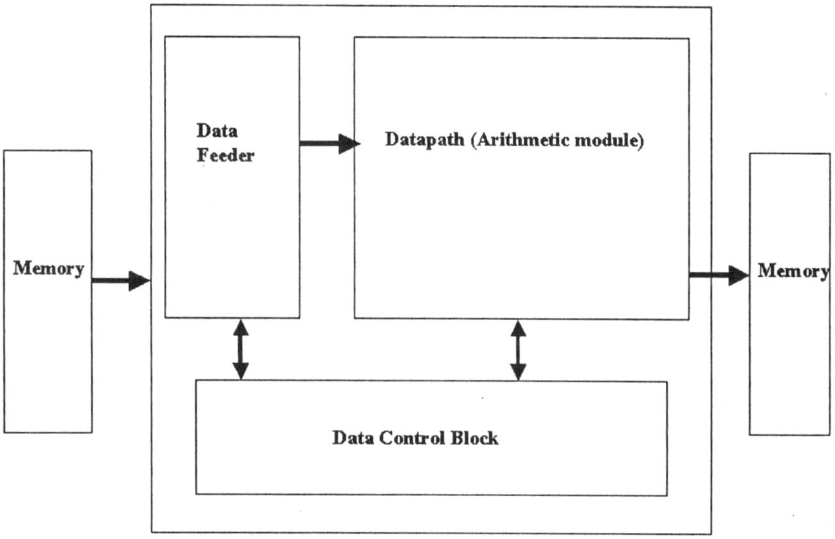

Figure 4-5. Block diagram of JPEG model

4.3.1 A closer look at the individual modules

Data feeder module

Figure 4-6 shows a block diagram of the data feeder module. This module consists of two modules, a serial in parallel out module (SIPO) and a parallel in parallel out module (PIPO). The SIPO reads in one 16-bit data at a time and provides 64 16-bit data in parallel as output.

When enabled (sipo_enable), the SIPO will start pushing the input data into the shift registers. Once we have 64 data samples, the SIPO will be disabled and the PIPO will latch the valid data out. The latched data is used by the datapath module for further processing. The data control block generates the enable signals for the SIPO and PIPO. Figure 4-7 shows a sample waveform that shows the functionality of the data feeder module.

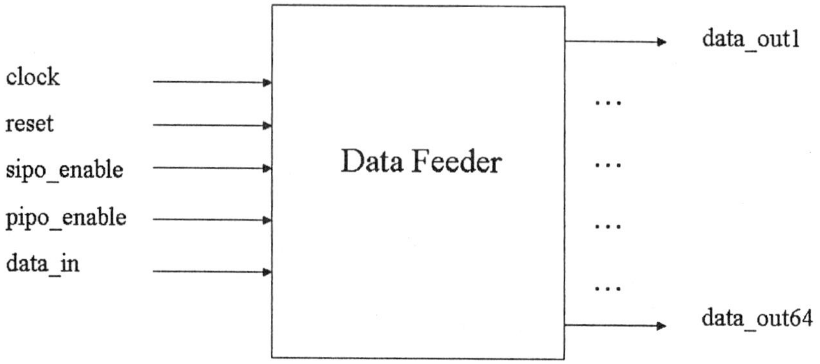

Figure 4-6. Data feeder block diagram

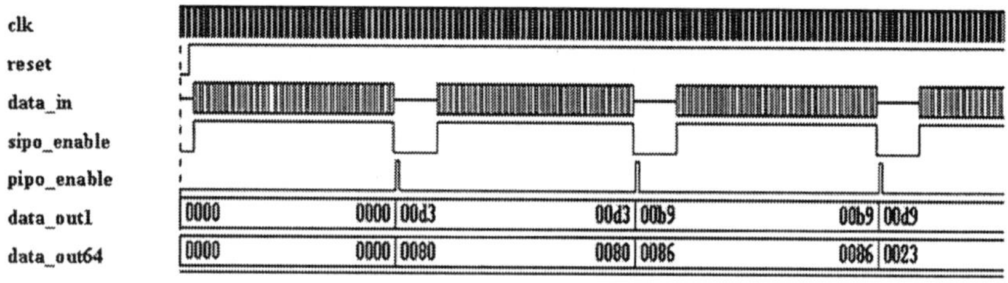

Figure 4-7. Waveform for Data feeder module

Data path module

The data path module takes in 64 16-bit data at a time and performs certain arithmetic operations on them. The process extends over multiple cycles to accommodate the completion of all operations. A multi-level pipeline is used to accomplish this task. Figure 4-8 shows a block diagram of the pipeline used to perform the arithmetic operations.

The data path is a simple latch based pipeline design. The data goes through four stages of processing, transform1, transpose, transform2 and quantization. After each stage of processing, the stable values are latched to the next stage by using a PIPO. The PIPO is a latch that is controlled by the enable signals generated by the data control module. Each module gets 2 clock cycles to complete their process. In other words, the data control module generates 4 enable signals at an interval of 2 clock cycles that are used to latch the stable data from the output of each stage. Figure 4-9 shows the relationship between the control signals of the pipeline.

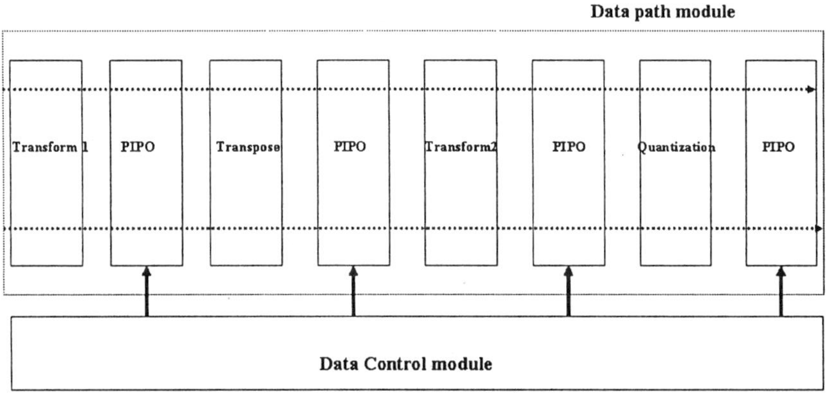

Figure 4-8. Block diagram showing details of the pipeline

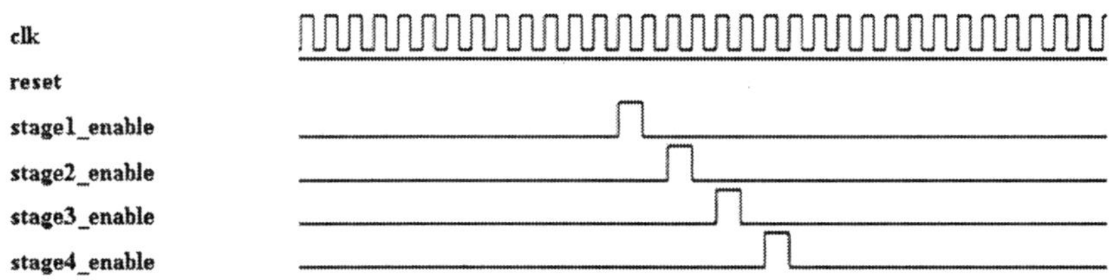

Figure 4-9. Waveform for pipeline control

Data control module

The data control module is a simple finite state machine (FSM). It produces the control signals required to keep the data moving along the pipeline smoothly. Figure 4-10 is a sample block diagram of the data control module and Figure 4-11 is a waveform showing the generation of the control signals.

The data control module generates the control signals for the data feeder and the data path modules. The state machine starts operating once it gets a "get_data" signal from outside. This kicks off the counters to generate the "read" signal for the memory and also the "read_address." It helps read 64 valid data every time.

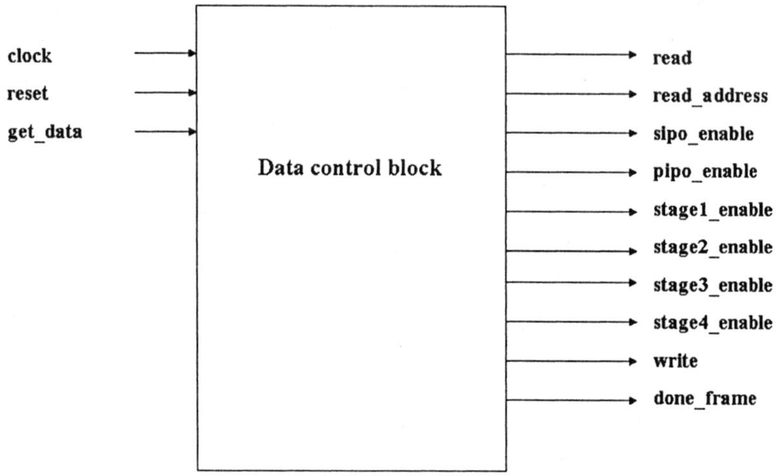

Figure 4-10. Block diagram for data control block

Once 64 valid data is read, the "read" is disabled and the enable signals for the pipeline are generated sequentially. After the data is processed through the 4 stages of pipeline, the "write" signal is generated to store the processed data into a memory model. After the "write," a fresh set of 64 data is read from the memory. This process continues until all the data is read from the memory. In the sample JPEG design used, the memory can hold 262144 bytes (equivalent to a 512 X 512 image). This means that the control signal generation is repeated 4096 (262144/64) times to finish the processing of all data points. After completing all blocks, the control block asserts the "done_frame" signal and immediately the "get_data" signal is de-asserted.

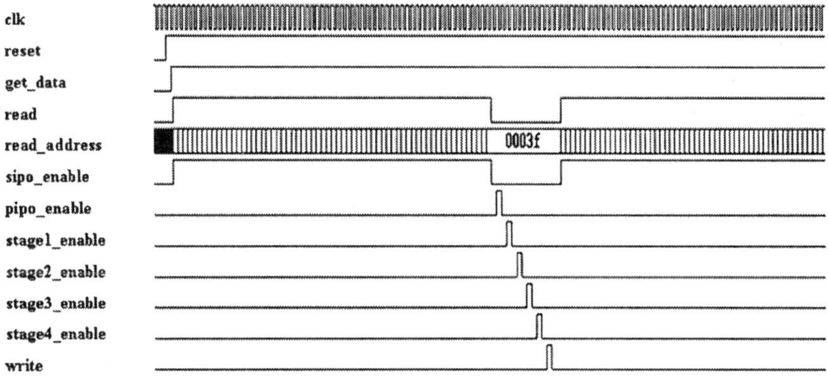

Figure 4-11. Waveform for control block

What needs to be verified?

- The validity of the data flow control signals. Are they generated correctly according to the timing requirements?
- Is the data path working correctly, is it producing valid output data to be stored in the memory?

4.3.2 SVA Checkers for the JPEG design

Based on the description of the design, the following list of checkers needs to be written to verify the design thoroughly.

JPEG_chk1: "get_data" and "done_frame" signals are mutually exclusive.

The design starts reading data from the memory when the "get_data" signal is asserted. While acquiring and processing data, "done_frame" signal should be held low. When all data has been processed, the design asserts the "done_frame" signal and de-asserts the "get_data" signal. Hence, these two signals can never be asserted at the same time.

```
property p_mutex;
  @(posedge clk) ((reset_) |->
              not (done_frame && get_data));
endproperty
a_mutex: assert property(p_mutex);
```

This is a mutually exclusive condition and can be written easily with a "**not**" operator. The "**not**" operator states that the test expression can never be true. The checker kicks off on every positive edge of the clock. The result of the checker "a_mutex" is shown in Figure 4-12.

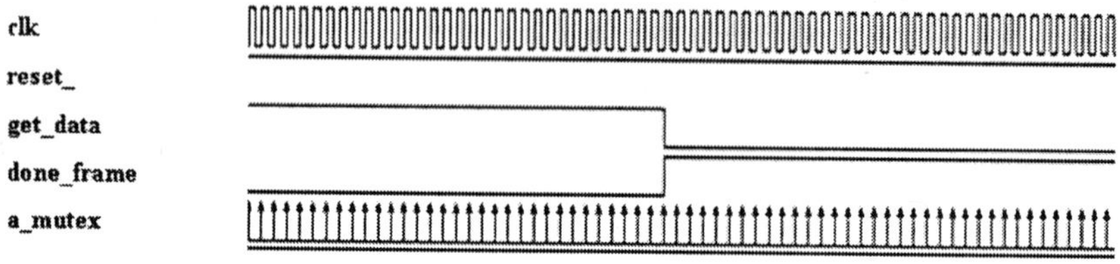

Figure 4-12. Waveform for JPEG_chk1

JPEG_chk2: The "read" signal is held high for 64 cycles continuously and during this period, the "read_address' is incremented by one in every clock cycle.

The sample design processes 64 data points at a time, which means that, a burst read is done for 64 cycles to get all the data that need to be processed. By verifying the above statement, we prove that we read a unique data on each of the 64 clock cycles.

```
sequence s_read;
  (rd_addr == $past (rd_addr)+1) [*0:$] ##1
                $fell (rd);
endsequence

property p_read;
  @(posedge clk)
((($rose (rd) && reset_) |->
                        s_read);
endproperty

a_read: assert property(p_read);
```

The read address is checked for the increment by using the **$past** operator. The value of "rd_addr" in the current clock cycle should be the value in the previous clock cycle incremented by 1. This checking is done from the rising edge of the read signal (**$rose** (rd)) until the falling edge of the read signal (**$fell** (rd)). The "**repeat until [*0:$]**" operator is used to check the validity of the address (until the "rd" signal is de-asserted). The result of the check "a_read" is shown in Figure 4-13. Every time there is a rising edge on the read signal, a valid match on the property is shown. The property itself will be active for several clock cycles but the match is indicated only once in the results and this is the point where the property begins to become active.

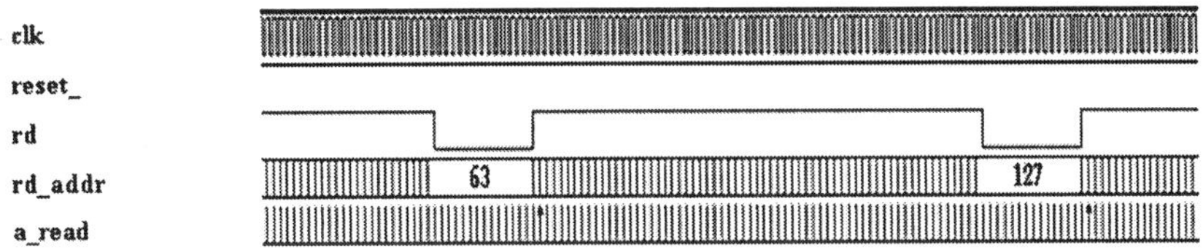

Figure 4-13. Waveform for JPEG_chk2

JPEG_chk3: The "sipo_en" is held high for 64 cycles during the read cycle and then disabled, 2 cycles later the "pipo_en" signal is asserted to latch the data that will be processed by the datapath module.

The data feeder module depends on the "sipo_en" and the "pipo_en" signals to provide the valid data to the datapath module. This checker verifies the functionality of the data feeder module.

```
sequence s_datafeeder;
    sipo_en[*64] ##1 $fell (sipo_en) ##1
        latch_en ##1 !latch_en;
endsequence

property p_datafeeder;
 @(posedge clk) ($rose (sipo_en) && reset_) |->
        s_datafeeder;
endproperty

a_datafeeder: assert property(p_datafeeder);
```

A simple **repeat operator** (*) is used to monitor whether the "sipo_en" signal is held high for 64 clock cycles. Once the "sipo_en" signal goes down, a "latch_en" pulse is asserted. The result of the check "a_datafeeder" is shown in Figure 4-14.

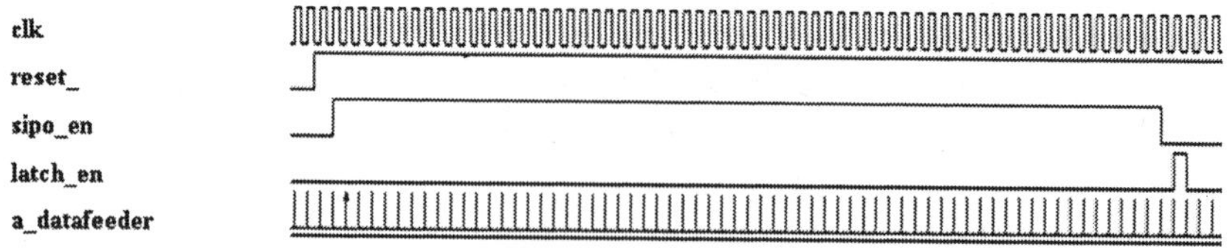

Figure 4-14. Waveform for JPEG_chk3

The sequence "s_datafeeder" can also be written as follows.

```
sequence s_datafeeder;
   ##64 $fell (sipo_en) ##2
   latch_en ##1 latch_en;
endsequence
```

In this description of "s_datafeeder," the falling edge of "sipo_en" is checked after 64 clock cycles and it does not guarantee that "sipo_en" was held high during these 64 clock cycles.

JPEG_chk4: In the datapath module, each stage is enabled with a gap of 2 clock cycles.

Every stage in the data path has 2 clock cycles to provide the stable value for the next stage. The signals "dp1_enable," "dp2_enable," "dp3_enable" and "dp4_enable" help latch the stable data at each stage and they are asserted in a sequence of 2 clock cycle gaps. This makes sure that the data flow is happening correctly.

```
sequence s_control;
 dp1_en ##1 !dp1_en ##1 dp2_en ##1 !dp2_en ##1
 dp3_en ##1 !dp3_en ##1 dp4_en ##1 !dp4_en ##1
 wr ##1 !wr;
endsequence

property p_control;
 @(posedge clk) $fell (latch_en) |=> s_control;
endproperty
a_control: assert property(p_control);
```

Each enable signal for the PIPO is a pulse of one clock cycle and they are generated 2 clock cycles apart. The sequence "s_control" monitors the rise and fall of each one of these control signals in a simple sequential concatenation method. The sequence starts when the data is loaded into the data path module (latch_en). An **implication operator** (**|->**) is used to indicate that the falling edge of the signal "latch_en" is the gating condition for the rest of the sequence to be tested. The sequence ends when the data has passed through the data path and the processed data has been written into the memory (wr).

JPEG_chk5: From the rising edge of "get_data" to the falling edge of "get_data," sequences "s_datafeeder" and "s_control" are repeated 4096 times unless the design is reset.

This guarantees that all data has been processed in the correct order. This is also used as a functional coverage check to make sure that all data from the memory has been processed.

```
property p_control_all;
  @(posedge clk) ($rose (sipo_en) && reset_) |->
        s_datafeeder ##1 s_control;
endproperty

property p_block;
 @(posedge clk)
  $fell (get_data) && $rose(done_frame) |->
                          (block == 4095);
endproperty

a_control_all: assert property(p_control_all);
c_control_all:
     cover property(p_control_all)block++;

a_block: assert property(p_block);
```

To make sure that the entire sequence repeats 4096 times to process all the data points, the checks JPEG_chk3 and JPEG_chk4 should be concatenated. The property p_control_all will start when the data is read from the memory (sipo_en) and will end when the processed data has been written to the output memory (wr). A "cover" statement can be declared for the property p_control_all that will provide information on how many times the property really succeeded and how many times it succeeded vacuously. If there is a real success, the variable "block" is incremented by one each time. The number of real successes should equal 4095. The property a_block uses this variable to verify that all blocks of data have been verified. If the "done_frame" signal has a rising edge and on the same clock cycle if the "get_data" signal has a falling edge, that indicates that the last block of data is being processed and at this point the variable "block" should indicate a value of 4095. The result of the check "a_control" is shown in Figure 4-15. The result of the check "a_block" is shown in Figure 4-16.

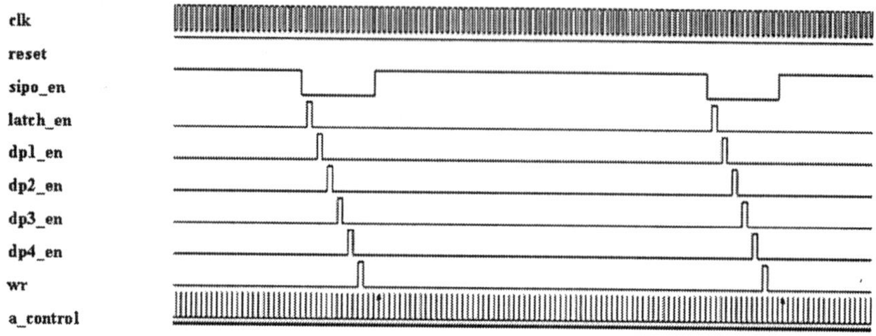

Figure 4-15. Waveform for JPEG_chk5

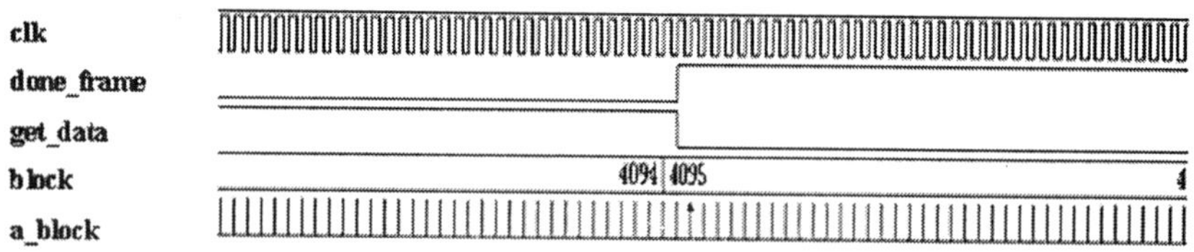

Figure 4-16. Waveform for check a_block

4.3.3 Data checking for the JPEG model

In Section 4.3.2, SVA checkers were written to verify that all the control signals are generated correctly. This guarantees that the data is moving along the pipeline smoothly. This does not check for data integrity. Each block in the pipeline performs some transformation to the data and this needs to be verified. A very common method used to verify the data is by dumping the output to a file. This output file is later compared with the result produced by the golden model as a post-process. While this method could work, it has a few disadvantages:

1. The simulation has to finish in order to compare the output with the expected results. If the output is wrong, then a lot of simulation cycles have been wasted.

2. This method is not very debug friendly since it does not say at which stage of the pipeline the output data started failing. One simple way to overcome this would be to dump the output of each and every pipeline stage and do the comparisons. While this will help debug, it will still waste simulation cycles as mentioned before..

A more efficient way to do the data checking will be to do the comparison dynamically. This can be accomplished in several ways and each user has to decide which one is good for his or her simulation environment. Since the data checking is a repetitive process, the dynamic comparisons can be shut down after a few data packets have been verified. In other words, a goal can be set to gain confidence on the dynamic data checking process and once the goal is attained, these checkers can be shut down, hence improving simulation throughput.

Possible steps for dynamic data checking of JPEG model:

- Simulate the golden C model that will produce results for each pipeline stage as shown in Figure 4-17. While it is not always easy to match the RTL pipeline stages with that of the golden C model, it is also not impossible.
- Generate the following output data files from the golden C model of the JPEG design before simulating the actual RTL - Wh1.dat (output of transform1), Xpose.dat (output of transpose), Wh2.dat (output of transform1), Quantize.dat (output of quantization).
- Use the same input data file on the RTL to perform data checking.
- Create a generic checker that can load the golden results into the simulation environment and then compare them dynamically as the RTL simulates, as shown in Figure 4-18. As the simulation proceeds, at the relevant trigger points, the checker will compare the golden results with the design results and report any failures. A sample data checker is shown in Example 4.3.

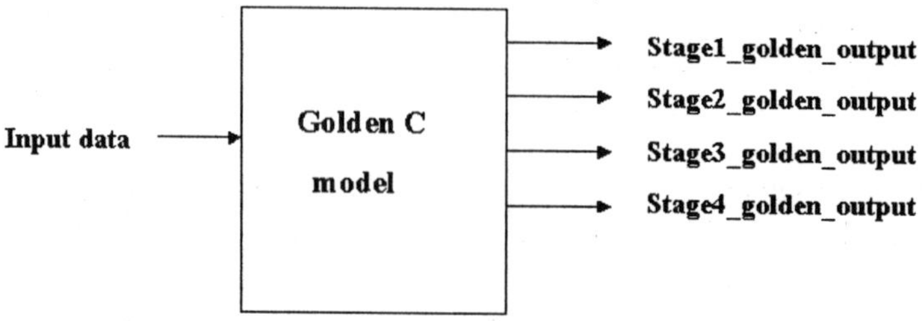

Figure 4-17. Golden output from C model

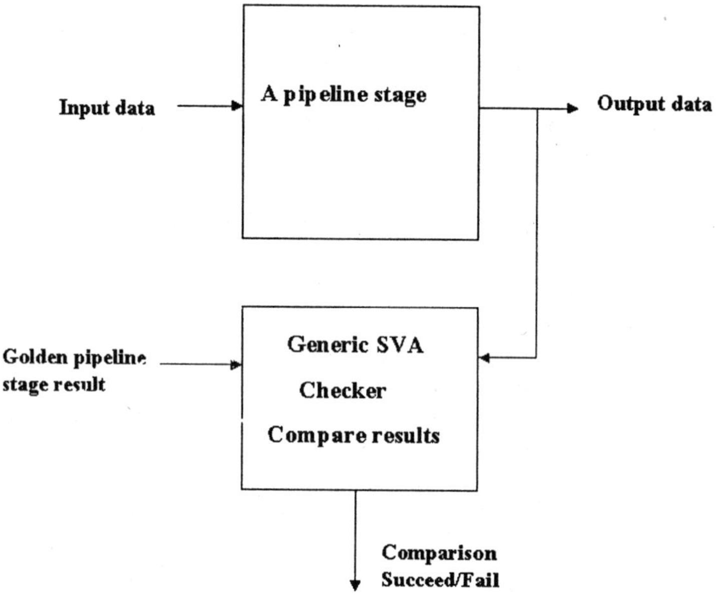

Figure 4-18. Dynamic Pipeline checker

Example 4.3 SVA checker for data-path verification

```
module dp_chk(
input logic reset, clk, enable,
input logic [15:0]
      d1, d2, d3, …, d61, d62, d63, d64);
parameter data_file = "";
parameter identity = "";
```

```
integer i=0, j=0;
integer blk=0;
integer fd, fdl;

logic [31:0] pix_in_temp;
logic [15:0] local_array[0:63];
logic [15:0] pix_in [0:262143];

// use $fopen construct to open the golden
// results file

initial
begin
fd = $fopen(data_file, "r");
end

// copy design data to a local array

always@(*)
begin
local_array[0] <= d1;
local_array[1] <= d2;
local_array[2] <= d3;
..
..
local_array[62] <= d63;
local_array[63] <= d64;
end

// load actual results

always@(negedge enable)
begin
  if(reset)
  $display
  ("\nDATA CHECKING: Block number %0d\n", blk);
  for(j=0; j<64; j++)
  begin
  fdl = $fscanf(fd, " %x", pix_in[j]);
  end
  blk++;
end
```

```
// compare results

genvar k;
generate
for(k=0; k<64; k++)
begin: dchk
a_dp_chk: assert property(
  @(posedge clk) (reset && $fell(enable)) |=>
  (pix_in[k] == local_array[k])) else $fatal;
end
endgenerate

endmodule

// check that data is put into blocks of 64
correctly

bind data_feeder dp_chk
#(.data_file("input_image.dat"),
.identity("INPUT")) dpchk1
(reset_, clk, latch_en, q0, q1, ....,
q61, q62, q63);

// check that the output of first wh transform is
// correct

bind datapath dp_chk
#(.data_file("wh1.dat"), .identity("WH1"))
dpchk2
reset, clk, dp_enable1,
dwl1,dwl2,....,dwl61,dwl62,dwl63,dwl64);

// check that the transposed data is correct

bind datapath dp_chk
#(.data_file("xposed.dat"),
.identity("TRANSPOSE")) dpchk3
 (reset, clk, dp_enable2, dwlt1, dwlt2, ....,
 dwlt61, dwlt62, dwlt63, dwlt64);

// check that the output of the second wh
// transform is correct
```

```
bind datapath dp_chk
#(.data_file("wh2.dat"), .identity("WH2"))
dpchk4
  (reset, clk, dp_enable3, dwltwl1, dwltwl2, ....,
 dwltwl63, dwltwl64);

// check that the output of quantization is
// correct

bind datapath dp_chk
#(.data_file("quantized.dat"),
.identity("QUANTIZATION")) dpchk5
(reset, clk, dp_enable4,
do1,do2,....do62,do63,do64);
```

Example 4.3 shows a generic SVA datapath checker and how it is bound to the various stages of the pipeline design.

- The checker defines 2 parameters that help identify the checker to be a unique one. The parameter "data_file" defines which golden file should be used by a specific instance of the checker. The parameter "identity" defines which section of the data path the checker is bound to.
- The golden data is stored in a file. This data file is opened for reading purpose using the $fopen construct.
- On trigger (the enable signal), the actual design outputs are stored in the checker locally (local_array). Note that the datapath processes 64 data points at a time and hence, only 64 data points should be read from the golden file on a trigger. A variable is incremented by 1 on each trigger to document which block of the image is being verified currently.
- Using a "generate" statement, 64 checkers are created, one for each data point. The "for" loop helps loop around and check all 64 data points on every trigger.
- The action block of the assert statement uses a $fatal construct. This instructs the simulator to exit the simulation if there is a violation. This prevents running the simulation unnecessarily after finding mismatches.
- The checker can be connected to specific points of the data path by using the "bind" construct. By defining the parameters relevant to each point, each checker becomes a unique instance.

A sample simulation log is shown below.

```
DATA CHECKING: Block number 1
"adv_datapath.sva", 118:
tb.jpeg_int.datapath_inst.dpchk5.dchk[0].a_dp_ch
k: started at 795s failed at 805s
 Offending '(pix_in[0] == local_array[0])'
"adv_datapath.sva", 118:
Fatal: "adv_datapath.sva", 118:
tb.jpeg_int.datapath_inst.dpchk5.dchk[0].a_dp_chk
: at time 805
```

Note that the failure is coming from the instance of the checker attached to the "QUANTIZATION" module (instance dpchk5) of the data path. The failure clearly points out the time of failure and which data of the block failed. For example, in the above log, data point 0 of Block 1 failed.

While this is one way to perform dynamic data checking, this might not be suitable for all designs. Each design is different and they have different specifications and requirements. This method can be used as a model to derive a methodology suitable for a specific design.

4.4 Summary for data intensive designs

- SVA provides the capability to perform arithmetic operations and is capable of using most SystemVerilog data types.
- By using Verilog tasks and functions, data checking can be automated and functional coverage information on the design can be obtained.
- Dynamic SVA checkers for data path uses the simulation cycles wisely and does not wait until the end of the simulation to find about design problems. The checkers also make debugging easy by pointing to the exact area of failure.

Chapter 5

SVA FOR MEMORIES

Memory controller protocol

Computers and consumer electronic devices have huge amount of memory to store multimedia application data. In the ASIC world, almost all the chips that are being designed today use embedded memory (DRAM, SRAM, ROM, etc.). As the memory access time is becoming faster and faster, it becomes very essential that the end product work with multiple memory vendors and with different timing requirements. The major bottleneck in the verification of memory controller interface is the timing of the control signals. This can be effectively done using assertions. The assertions written for a particular type of memory device can be re-used with multiple vendors just by modifying the timing parameters. This chapter discusses developing reusable SVA checkers for different types of memory devices.

5.1 Sample System – Memory controller

The sample system has a CPU that interfaces with a memory controller. The CPU can read and write data to the various memories connected to the memory controller. The memory controller can interface to different type of memories such as SDRAM, DDR-SDRAM, SRAM, Flash, ROM, etc. The block diagram for the sample system is shown in Figure 5-1.

5.1.1 CPU - AHB Interface Operation

The CPU is a generic processor that uses the AHB bus interface to interact with the memory controller. The CPU generates the read/write commands. The CPU also generates the chip select signals for selecting the memory with which the read/write operation will be performed. It supports both separate and shared memory address/data busses to SDRAM and static memories. It supports the AHB data width of 32, 64 or 128 bits. It also supports the AHB 32bit wide address bus.

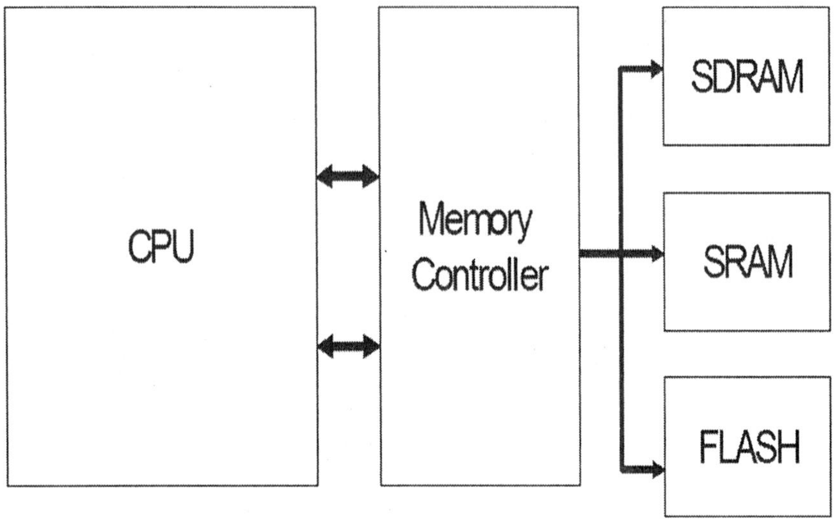

Figure 5-1. System block diagram

Figure 5-2 shows the signal interface between the CPU-AHB bus interface and the memory controller. A brief description of the pins is listed below.

- hclk – clock generated by the CPU
- haddr – read/write address generated by the CPU
- hwdata – write data generated by the CPU
- hrdata – read data to the CPU from the memory
- hready – ready signal from CPU
- hready_resp – memory acknowledge signal for hready
- hsize – configures the size of the data transfer from CPU to memory
  - 00 – 128 bits of data transfer at a time
  - 01 – 64 bits of data transfer
  - 10 – 32 bits of data transfer
- hburst – defines how the memory address is accessed
  - 000 – single – one single memory location
  - 001 – INCR – Increments address 0x40, 0x44 …

- 010 – WRAP4 – Wraps address in 4 word boundaries (0x48, 0x4c, 0x40 0x44)
- 011 – INCR4 – Increments address by 4 words from the current address
- 100 – WRAP8 – Wraps address in 8 word boundaries
- 101 – INCR8 – Increments address in 8 word blocks
- 110 – WRAP16 – Wraps address in 16 word boundaries
- 111 – INCR16 – Increments address in 16 word blocks

- sel_mem, sel_reg – select whether to do transaction with the external memories or the registers

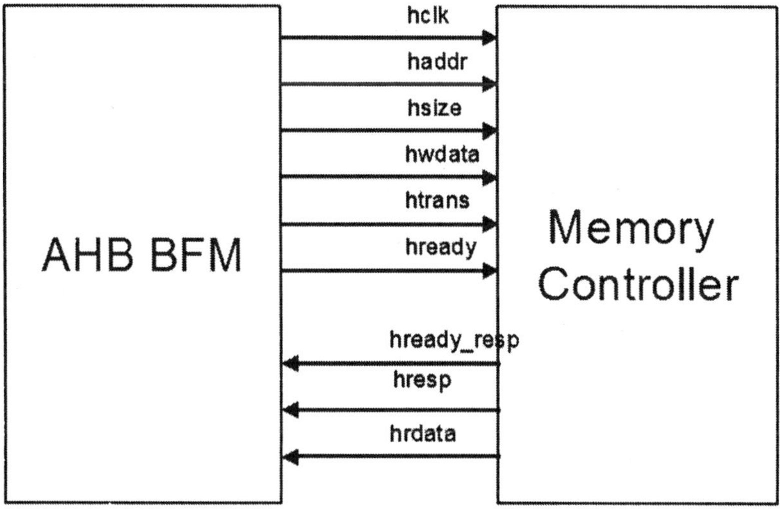

Figure 5-2. CPU block diagram

Figure 5-3 shows the waveform of a CPU write transaction to the memory controller. The CPU initiates a write transaction to the memory controller at marker 1. The size (hsize) is set to "10" and hence the data transaction is 32 bits wide. The burst (hburst) is set to "010" and hence the burst type is WRAP4. Based on the burst type, the address access will be 0x0, 0x4, 0x8 and 0xc. Hence, the write transaction of size 32 bits and of type WRAP4 is initiated at marker 1. The figure shows that the address is incrementing from 80000000 to 8000000c and the data on "hwdata" is being written to the SDRAM.

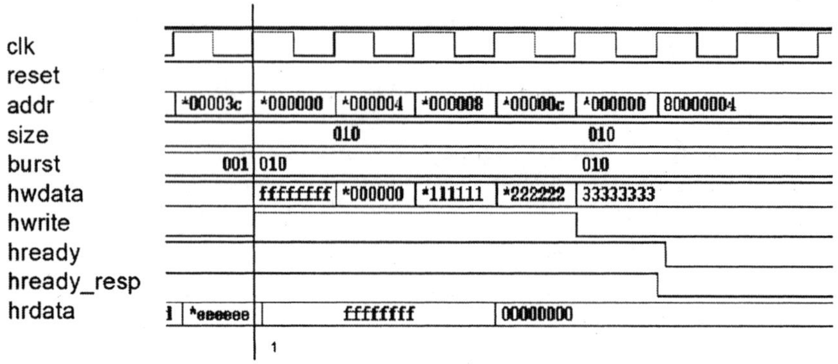

Figure 5-3. CPU-AHB write

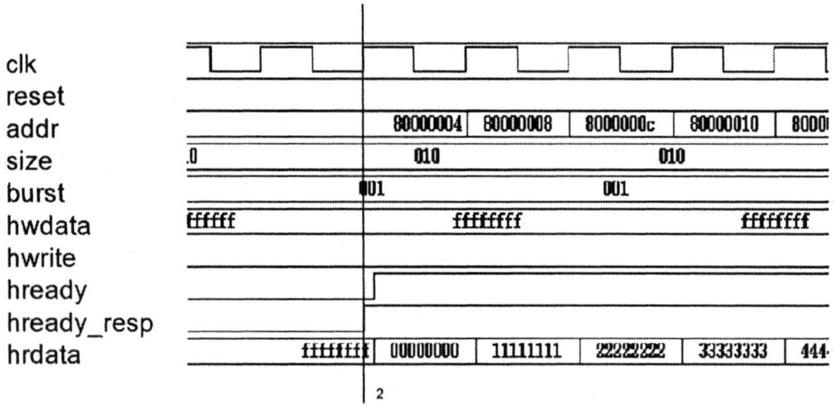

Figure 5-4. CPU-AHB read

Figure 5-4 shows the waveform of a CPU read transaction to the memory controller. A read transaction of size 32 bits and type WRAP4 is initiated at marker 2 when the "ready" and "ready_response" signals are asserted. The figure shows that the same data that was written is being read.

5.1.2 Memory controller operation

The memory controller in the sample system interfaces to the SDRAM, SRAM and Synchronous Flash devices. A block diagram of the connection is shown in Figure 5-5.

SDRAM Interface

The memory controller interface to the SDRAM is generic. The interface is fully synchronous and all signals are registered on the positive edge of the clock. Read and write access to the SDRAM is burst oriented and they start at the address specified by the AHB bus and continue for a programmed number of locations. The connection from the memory controller to the SDRAM is direct and has no glue. It supports 16 SDRAM address bits. It also has programmable row and column address widths. All the SDRAM timing parameters are programmable. It supports auto-refresh with programmable refresh intervals.

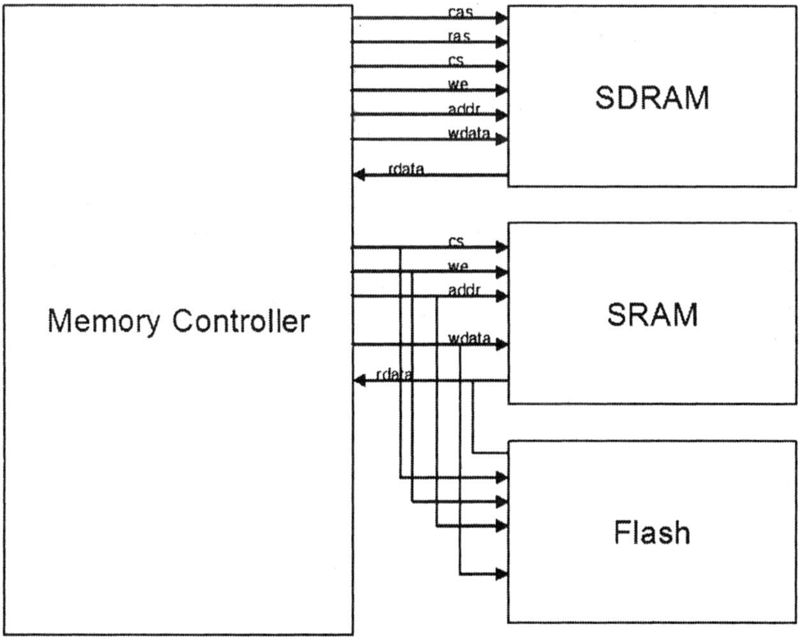

Figure 5-5. Memory Controller Block Diagram

A brief description of the pin connections between the SDRAM and the memory controller is listed below.

- clk – clock input to SDRAM
- ras_ – selects the row address

- cas_ – selects the column address
- we_ – when asserted write signal, when de-asserted read signal
- sel_ – when asserted SDRAM is selected
- data – bi-directional data bus for both reads and writes
- addr – read/write address
- bank_sel – selects a particular bank of SDRAM

A sample waveform for a write command issued by the memory controller is shown in Figure 5-6. A burst write of length four is written and read back. An active command (ras and chip select are asserted) is issued and then a write command (cas, we, and sel are asserted) is issued to the address 0x0000. Figure 5-6 shows that the data is written with a burst length of 4 to the memory at marker 1. Figure 5-7 shows a burst read command issued by the memory controller. The data is read out of the memory (marker 1) after a "cas" latency of two clock cycles.

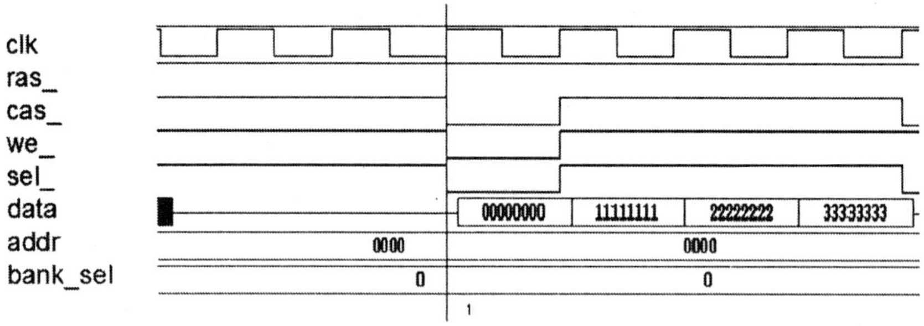

Figure 5-6. SDRAM write operation

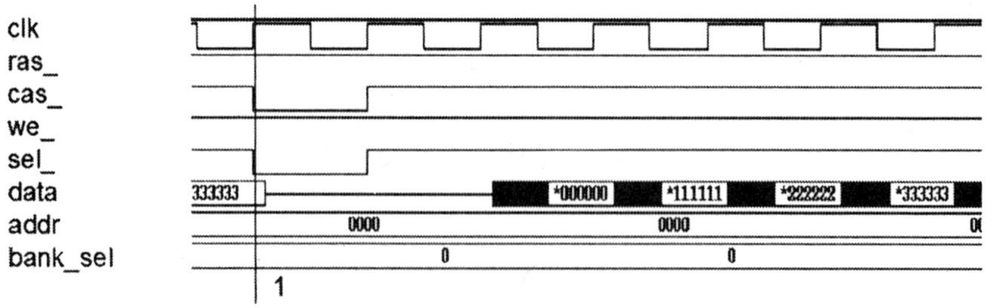

Figure 5-7. SDRAM read operation

SRAM/FLASH Interface

The SRAM/FLASH memories are static memories and the interfaces are similar. The memory controller supports asynchronous SRAMs, page-mode flashes and ROMs. The address width can be configured up to 32 bits. It also has the "ready" handshake signal to support non-SRAM type devices. The memory data width can be configured to 8 bits, 16 bits, 32 bits, 64 bits or 128 bits. The static memory data width can be a minimum of 8 bits instead of the 16 bits standard requirement. The flash memory used in the sample system is write protected, so that the important system information is protected in the boot block.

A brief description of the pin connections between the SRAM and the memory controller is listed below:

- addr – address pins to the static memories from memory controller
- data – data to/from the static memories to the memory controller
- sel_ – chip select pins to select the corresponding static memory
- we_ – write when asserted
- oe_ – output enable asserted during read
- bs_ – byte control pins to enable different data widths

The interface for flash is similar to SRAM except that it has two more signals:

- wp_ – write protect pin
- rp_ – reset power down pin

Figure 5-8 shows a sample waveform for the interface between the memory controller and the SRAM. When signals "we_" and "sel_" are asserted (marker 1), a write is done to the SRAM. Similarly, when signals "sel_" and "oe_" are asserted (marker 2) and signal "we_" is de-asserted, a read is done from the SRAM. Figure 5-9 shows a sample waveform for the interface between the memory controller and the flash memory. The figure shows a burst read from the flash device. When signals "sel_" and "oe_" are asserted, a read operation from the address specified in the address bus (addr) is performed.

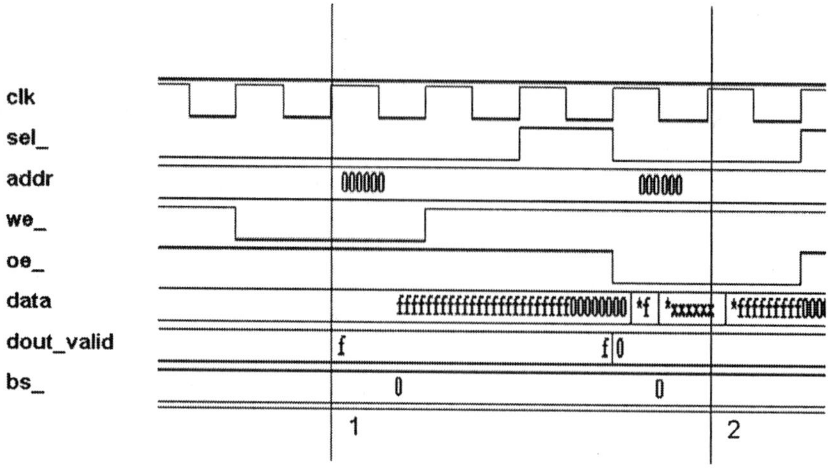

Figure 5-8. SRAM interface signals

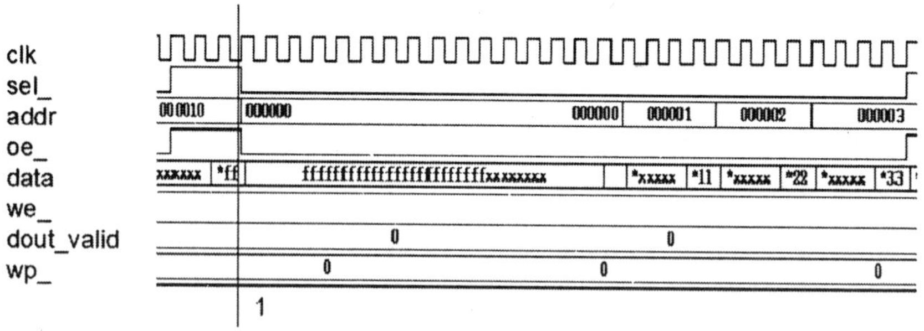

Figure 5-9. Flash interface signals

5.2 SDRAM Verification

This section will discuss how to verify the SDRAM control signals. There are a lot of timing parameters for SDRAM device and assertion based verification can be used effectively to verify that these timing requirements are not violated. The sample system uses the following SDRAM configuration.

512Mb SDRAM - 8M X 16bit X 4 bank

The 512Mb SDRAM under verification is a quad bank SDRAM and includes a synchronous interface. All signals are registered on the positive edge of the clock. Each of the 4 banks is organized as 8192 rows X 1024 columns X 16 bits. Read and write access to the SDRAM is burst oriented. The access starts at a selected location and continues for a programmed number of locations. Read/Write access always begins with an active command followed by a read/write command. The address bit corresponding to the active command denotes the row address and the bank that is selected. A0-A11 denotes the address and BA[1:0] denotes the bank that is being accessed. The address bits corresponding to the read/write command denote the starting column address (denoted by A0-A7).

The different combinations of the SDRAM interface signals sel_, ras_, cas_ and we_ constitute the different commands. All the SDRAM commands are summarized in Table 5-1. The "Command Inhibit" condition prevents the SDRAM from executing the new commands, regardless of whether the clock signal is enabled or not. Operations already in progress will not get affected (sel_ = 1, cas_, ras_, we_ = x).

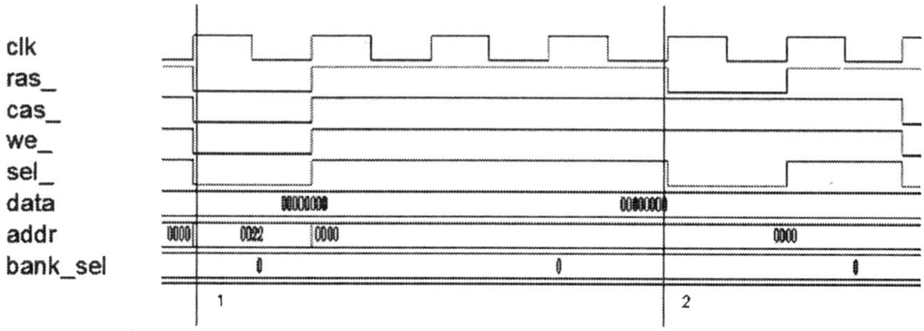

Figure 5-10. Load Mode Register/Active command

- **No Operation**: This prevents unwanted commands from being registered in idle/wait state (sel_ = 0, cas_, ras_, we_ = 1).

- **Load Mode Register**: The register is loaded through the address bus (A0-A11). The Load mode register is issued only when all the banks are idle (sel_, cas_, ras_, we_ = 0). Figure 5-10 shows a "load mode register" operation at marker 1 and an "active" operation at marker 2.

- **Active**: This command is issued to activate/open a row for access. The value on the address bus is the value of the row and the value on the bank_address bus specifies the bank (sel_, ras_ = 0; cas_, we_ = 1).

Table 5-1. SDRAM Commands

Command	Ras	Cas	We	Sel
No Operation	H	H	H	L
Active	L	H	H	L
Read	H	L	H	L
Write	H	L	L	L
Burst Terminate	H	H	L	L
Load Mode Register	L	L	L	L
Precharge	L	H	L	L
Auto-Refresh	L	L	H	L

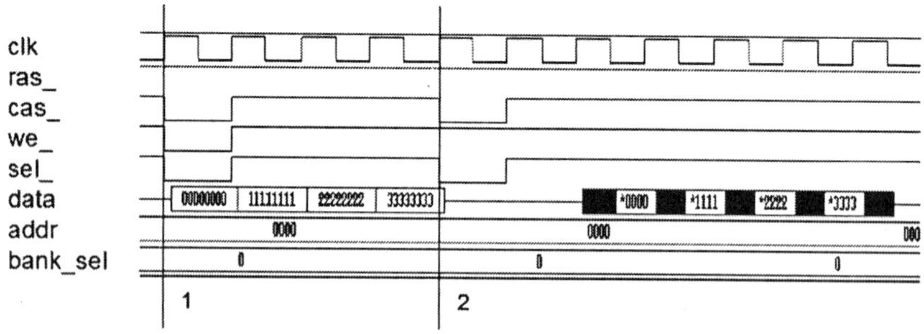

Figure 5-11. SDRAM read/write

- **Read**: This command is issued to do a burst read to an active row. The address provided on the bus "addr" provides the starting column address (sel_, cas_ = 0, ras_ we_ = 1).

- **Write**: This command is issued to initiate a burst write access to an open row. The address on the bus "addr" provides the starting column address (sel_, cas_, we_ = 0, ras_=1). Figure 5-11 shows a simple SDRAM read/write operation. A burst write is performed with a burst size of 4 at marker 1. A burst read to the same address location is done at marker 2.

- **Precharge**: Precharge is used to de-activate the rows (sel_, ras_, we_ = 0, cas_=1). If during precharge the addr[10] bit is set to 1, then all the rows in the banks are de-activated.

- **Auto-refresh**: This command is issued in the normal operation of the SDRAM. This command must be issued every time a refresh is required. All active banks must be precharged prior to issuing an auto-refresh.

- **Burst Terminate**: A burst terminate command is used to terminate a burst read or a burst write command.

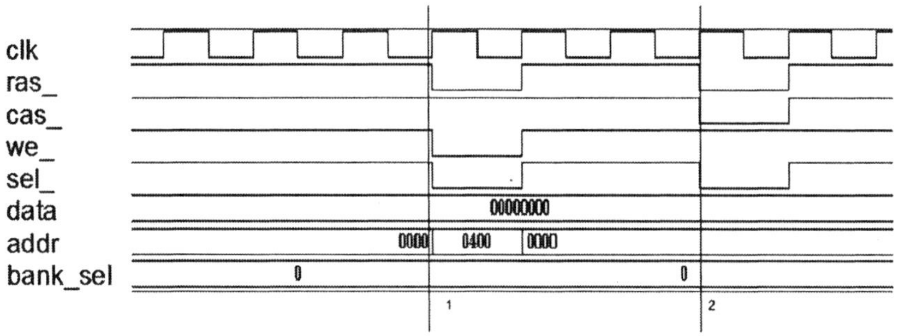

Figure 5-12. Precharge / Auto-refresh

Figure 5-12 shows the precharge command at marker 1 and auto-refresh command at marker 2.

A read/write operation to a SDRAM can be performed once the steps summarized in Figure 5-13 are completed. The description of the steps is as follows:

1. Initialization – once power is applied, SDRAM requires ~100us to initialize before any command can be issued.
2. Once the initialization is completed, one NOP/COMMAND INHIBIT is applied.
3. Then a precharge command is issued and all the rows are de-activated.
4. A Refresh command is issued after precharge.

5. Mode register is loaded (set the "cas" latency, burst size and other configurations).
6. Active command is issued (to activate the rows).
7. Read/Write command is issued.

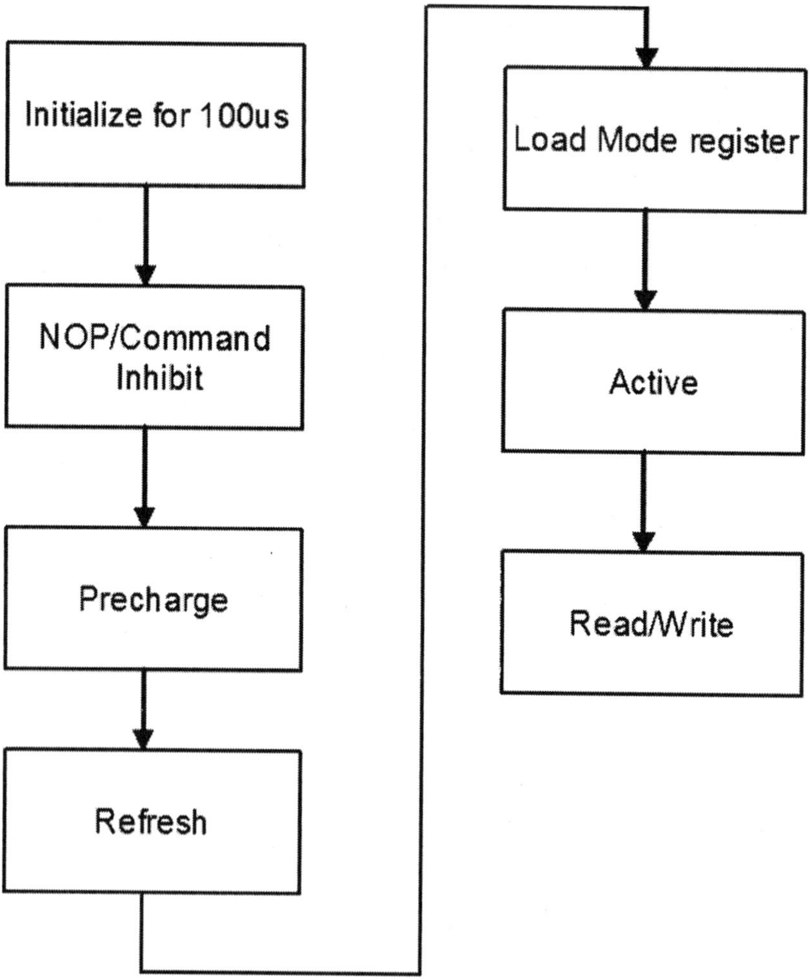

Figure 5-13. SDRAM operation flow chart

5.2.1 SDRAM Assertions

All the SDRAM commands like read, write, burst terminate, active, precharge, load mode register, etc. are derived from the four signals ras_, cas_, sel_ and write enable. Hence, all of these signals should be defined using `defines. These definitions can be re-used in the SVA checkers wherever necessary.

```
`define s_precharge
   (!ras_n && !sel_n[0] && !we_n && cas_n)

`define s_read
  (ras_n  &&  !sel_n[0]  &&  we_n  &&  !cas_n  &&
  (burst == 3'b000))

`define s_burst_read
   (ras_n  &&  !sel_n[0]  &&  we_n  &&  !cas_n  &&
   (burst != 3'b000))

`define s_write
   (ras_n && !sel_n[0] && !we_n && !cas_n)

`define s_autorefresh
   (!ras_n && !cas_n && !sel_n[0] && we_n)

`define s_loadmoderegister
   (!ras_n && !cas_n && !sel_n[0] && !we_n)

`define s_active
   (!ras_n && !sel_n[0] && cas_n && we_n)

`define s_write
    (!cas_n  &&  !we_n  &&  !sel_n[0]  &&  ras_n  &&
    (burst == 3'b000))

`define s_burst_write
   (!cas_n && !we_n && !sel_n[0] && ras_n &&
   (burst != 3'b000))
```

Some of the possible SVA checkers extracted based on the functionality of the SDRAM are shown below. The timing parameters used in these checkers specific to the SDRAM under consideration is listed in Table 5-2. Some of the timing parameters are specified in clock cycles and others in nanoseconds (ns). For the values specified in nanoseconds, the number of clock cycles is dependent on the clock cycle period (tCK). In this sample design the value of tCK is 10ns. Hence, the number of clock cycles is derived based on the value of the clock period and the value of the timing parameter provided in the specification of the SDRAM as shown below.

For example, tRCD=18ns

tRCD/tCK = 1.8 clock cycles

Hence, the timing window between an active command and a read/write command should be at least 2 clock cycles.

Table 5-2. Timing parameters for SDRAM

Parameter	Symbol	Min	Max
Load mode register to active	tMRD	4 cycles	4 cycles
Active to Active Command period	tRC	6 cycles (60ns)	-
Active to Read/Write	tRCD	2 cycles (18ns)	-
Read latency	tCAS	2 cycles	-
Auto Refresh period	tRFC	6 cycles (60ns)	-
Precharge Command period	tRP	2 cycles (18ns)	-
Active Bank a to Active Bank b	tRRD	2 cycles (12ns)	-
Active to Precharge command	tRAS	5 cycles (42ns)	12000 cycles (120000ns)

SDRAM_chk1: Load mode register to active command (tMRD).

The load mode register is used to load the mode register of the SDRAM with information on how the device is configured. Once the SDRAM is configured an active command should arrive in "tMRD" (4 clock cycles). Figure 5-14 shows that the load mode register command is sampled at marker 1 and four clock cycles later active command is sampled (marker 2), as expected. Hence, the check a_tMRD succeeds.

```
property p_tMRD;
     @(posedge clk)
     `s_loadmoderegister |->
                     ##[tMRD] `s_active;
endproperty
```

```
a_tMRD: assert property(p_tMRD);
c_tMRD: cover property(p_tMRD);
```

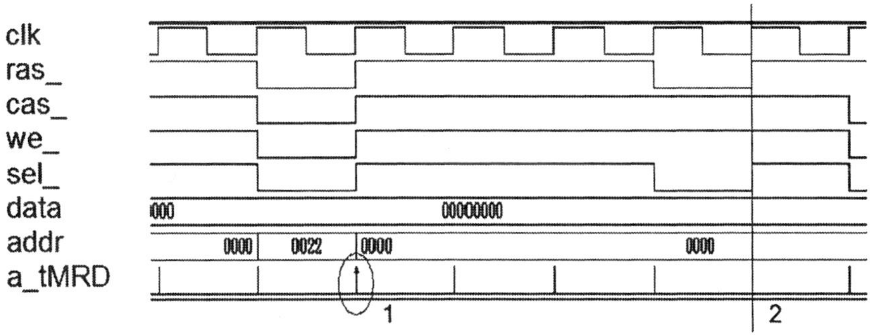

Figure 5-14. Load mode register to Active command, tMRD

SDRAM_chk2: Check the value of load mode register (022).

This assertion is used to check the value written into the mode register. This value is important as it determines the burst size and "cas" latency. When the load mode register command is issued, the value on the address bus is written into the load mode register. Bits [0:2] specify the burst size and the burst size is set to 4 (100). The "cas" latency value is set to 2. To set these parameters, the register has to be written with 0x0022.

Figure 5-14 shows a load mode register command being issued by the memory controller at marker 1. At this point, the address bus has a value of 0x0022. Hence, the check a_loadmoderegister succeeds.

```
property p_loadmoderegister;
        @(posedge clk)
        (`s_loadmoderegister) |->
                        (addr == 16'h0022);
endproperty
a_loadmoderegister:
        assert property(p_loadmoderegister);
c_loadmoderegister:
        cover property(p_loadmoderegister);
```

SDRAM_chk3: tCAS, read data is available with a latency of tCAS after the read command is issued.

In the sample SDRAM memory, whenever a read command is issued, the data is available after the "cas" latency (Column Address Select Latency). This is programmed in the mode register based on the memory vendor. Figure 5-15 shows that a read command is sampled at marker 1. After tCAS cycles, the data is valid as shown by marker 2. To verify this property, implication construct and **$isunknown** construct are used.

```
property p_read;
        @(posedge clk)
         (`s_read || `s_burst_read) |->
                  ##tCAS ($isunknowndata) == 0);
endproperty

a_read: assert property(p_read);
c_read: cover property(p_read);
```

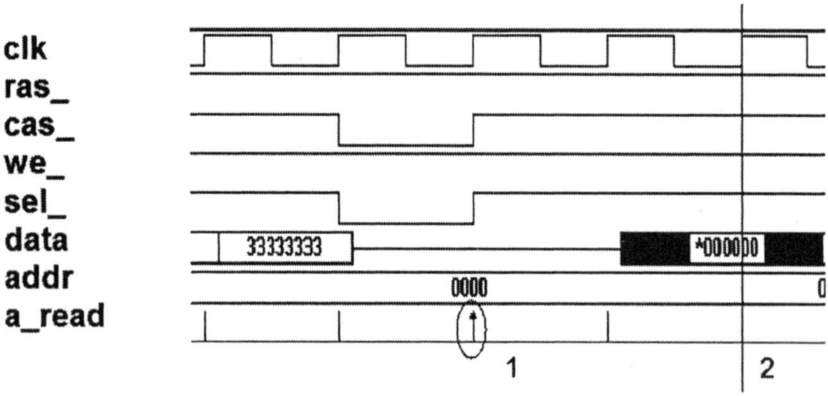

Figure 5-15. SDRAM read with tCAS latency

SDRAM_chk4: tRCD, after an active command, read/write can occur only after tRCD.

If the memory controller has issued an active command, then the read or write command cannot be issued within "tRCD" cycles. In the sample system used, once an active command is issued, a read/write command should be issued within 10 clock cycles. There are two specific conditions that need to be tested:

1. Once the active command is issued a read/write command does not occur within "tRCD" (this is a forbidden property).
2. Once the active command is issued, the read/write command must be issued within 10 clock cycles.

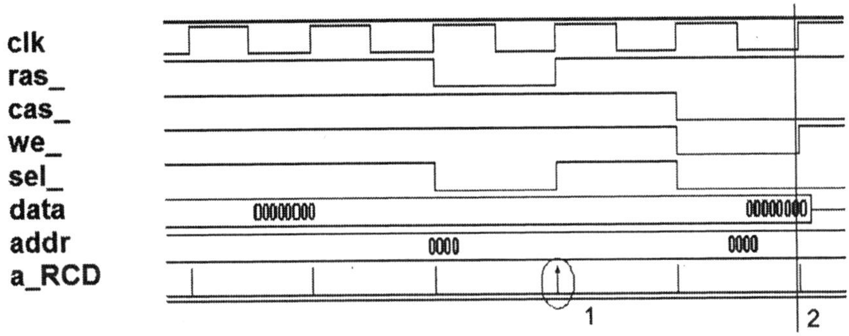

Figure 5-16. Active to Read/Write command, tRCD

```
property p_tRCD_not;
    @(posedge clk)
    `s_active |-> not ##[0: (tRCD - 1)]
          (`s_read || `s_write || `s_burst_read
          || `s_burst_write);
endproperty

property p_tRCD;
    @(posedge clk)
   `s_active |->
            ##[tRCD:10]  (`s_read  ||  `s_write  ||
           `s_burst_read || `s_burst_write);
endproperty

a_tRCD_not: assert property (p_tRCD);
a_tRCD: assert property (p_tRCD);

c_tRCD_not: cover property (p_tRCD);
c_tRCD: cover property (p_tRCD);
```

Figure 5-16 shows that the active command is sampled at marker 1. The write command is sampled 2 cycles after the active command at marker 2. Hence, the check a_tRCD is successful.

SDRAM_chk5: tRC, active to active command cannot come within tRC.

If an active command is issued, the controller cannot issue another active command within "tRC" (6 clock cycles). In the sample system used, if an active command is issued, then the next active command should be issued within 12000 clock cycles.

```
property p_tRC_not;
        @(posedge clk)
         `s_active |->
              not ##[1: (tRC - 1)] `s_active;
endproperty

property p_tRC;
        @(posedge clk)
         `s_active |->
               ##[tRC:12000] `s_active;
endproperty

a_tRC_not: assert property (p_tRC_not);
a_tRC: assert property (p_tRC);
c_tRC_not: cover property (p_tRC_not);
c_tRC: cover property (p_tRC);
```

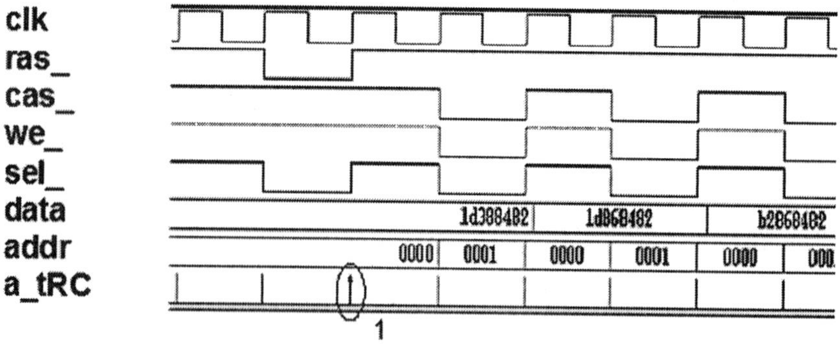

Figure 5-17. Active to Active command, tRC

Figure 5-17 shows the first active command with marker 1. The next active command arrives after 11,625 clock cycles (not shown in the figure) and hence the check a_tRC is successful.

SDRAM_chk6: tRFC, auto-refresh to auto-refresh cannot come within tRFC.

This property is similar to the previous property, in that the window between consecutive auto-refresh commands should be greater than tRFC. In the sample system used, if an auto-refresh command is issued then the next auto-refresh command should be issued within 12000 clock cycles.

```
property p_tRFC_not;
        @(posedge clk)
         `s_autorefresh |->
              not ##[1: (tRFC-1)] `s_autorefresh;
endproperty

property p_tRFC;
        @(posedge clk)
         `s_autorefresh |->
              ##[tRFC:12000] `s_autorefresh;
endproperty

a_tRFC_not: assert property (p_tRFC_not);
a_tRFC: assert property (p_tRFC);

c_tRFC_not: cover property (p_tRFC_not);
c_tRFC: cover property (p_tRFC);
```

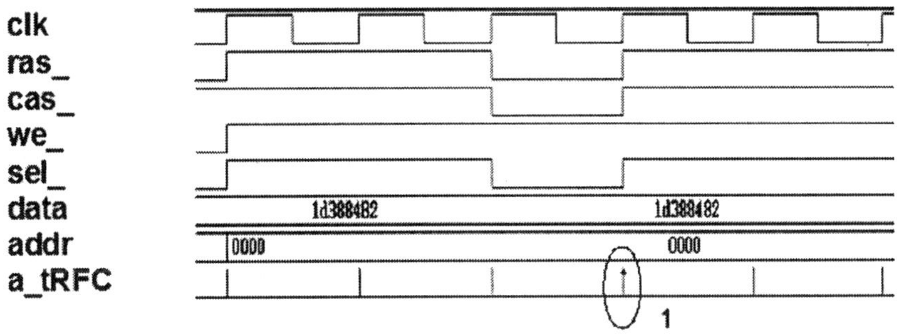

Figure 5-18. Auto-refresh to Auto-refresh command, tRFC

Figure 5-18 shows an auto-refresh command at marker 1. Another auto-refresh command arrives after 9 cycles (not shown in the figure). As this window is greater than "tRFC" as required by the memory specification, the assertion is successful.

SDRAM_chk7: Write command can follow a read command only after tCAS.

A write command cannot follow a read command immediately. By definition, a read command has a "cas" latency of 2 cycles. So the write can follow the read only after the "cas" latency window is satisfied.

```
property p_rd_wr;
        @(posedge clk)
         `s_read |->
                 not ##[0:tCAS] `s_write;
endproperty

a_rd_wr: assert property (p_rd_wr);
c_rd_wr: cover property (p_rd_wr);
```

SDRAM_chk8: tRP, precharge to active command cannot be issued until "tRP" is met.

The precharge command (de-activates the rows) to the active command (enables the rows) cannot happen within the "tRP" (2 cycles) window. In the sample system used, if a precharge command is issued, then an active command should be issued within 12000 clock cycles.

```
property p_tRP_not;
        @(posedge clk)
         `s_precharge |->
                   not ##[0:(tRP - 1)] `s_active;
endproperty

property p_tRP;
        @(posedge clk)
         `s_precharge |->
                   ##[tRP:12000] `s_active;
endproperty

a_tRP_not: assert property (p_tRP_not);
a_tRP: assert property (p_tRP);

c_trp_not: assert property (p_tRP_not);
```

```
c_trp: assert property (p_tRP);
```

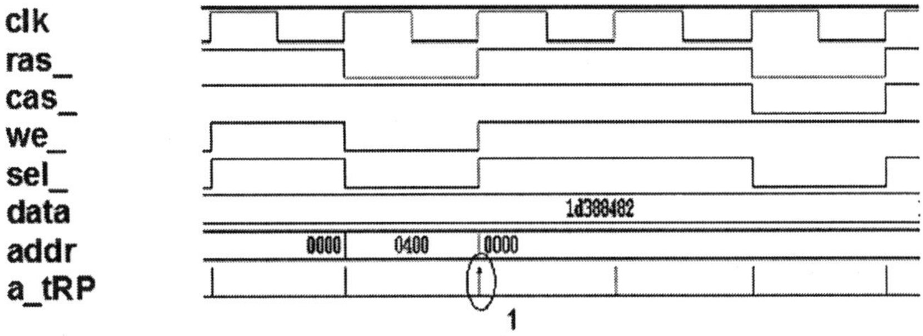

Figure 5-19. Precharge to Active command, tRP

Figure 5-19 shows a precharge occurring at marker 1. An active command is issued within 12000 cycles (not shown in figure) and hence the assertion is successful at marker 1.

SDRAM_chk9: tRAS, active to precharge must occur between tRASmin (5 clock cycles) to tRASmax (12000 clock cycles).

The active command (enables the rows) to the precharge command (de-activates the rows) cannot happen within the "tRASmin" cycles and should happen within "tRASmax" cycles.

```
property p_tRAS_not;
       @(posedge clk)
       `s_active |->
          not ##[0: (tRAS_min - 1)] `s_precharge;
endproperty

property p_Tras;
        @(posedge clk)
         `s_active |->
              ##[tRAS_min:tRAS_max] `s_precharge;
endproperty

a_tRAS_not: assert property (p_tRAS_not);
a_tRAS: assert property (p_tRAS);
c_tRAS: cover property (p_tRAS);
```

```
c_tRAS_not: cover property (p_tRAS_not);
```

SDRAM_chk10: Back to back writes are not allowed.

```
property p_wr_wr;
        @(posedge clk)
         `s_write |->
                     not ##1 `s_write;
endproperty

a_wr_wr: assert property (p_wr_wr);
c_wr_wr: cover property (p_wr_wr);
```

SDRAM_chk11: Check if auto-precharge is disabled during read/write operations.

Most of the SDRAM today can be precharged automatically by setting the addr[10] bit to a high during read/write operations. This assertion is written using implications and logical operation on the command definitions.

```
property p_disable_autoprecharge;
@(posedge clk)
(`s_write || `s_burst_write ||
`s_read || `s_burst_read) |->
                    addr[10] == 0;
endproperty

a_disable_autoprecharge:
       assert property(p_disable_autoprecharge);
```

Figure 5-20 shows the waveform for disabling auto-precharge during read/write commands.

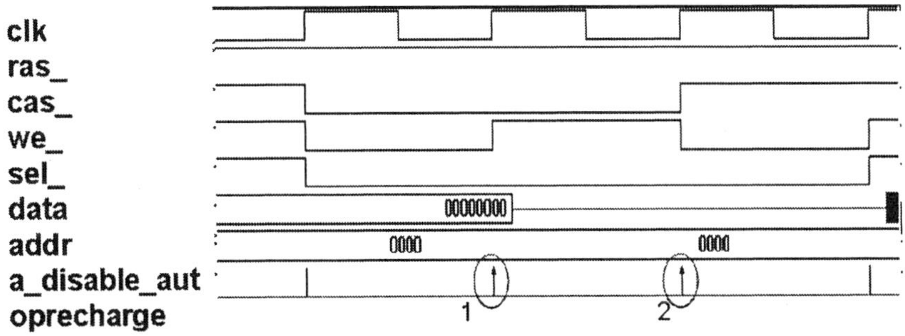

Figure 5-20. Disabling Auto-precharge

A write command is sampled at marker 1 and corresponding to the write, the addr[10] is 0. Similarly, when a read command is in progress (marker 2), the addr[10] is set to 0.

SDRAM_chk12: tRRD, minimum time interval between active commands to different banks is defined by tRRD (2 cycles).

Usually there are multiple banks in the SDRAM. There is a minimum time interval that is required between issuing active commands to different banks. The current system under verification has four banks.

```
property p_tRRD;
     @(posedge clk)
     (`s_active && bank_addr[1:0] == 0) |->
                 not ##[0: tRRD] (`s_active &&
                 bank_addr[1:0] != 0));
endproperty

a_tRRD: assert property(p_tRRD);
c_tRRD: cover property (p_tRRD);
```

This check verifies that if an active command is issued to bank 0, then an active command cannot be issued to other banks (1, 2, 3) within "tRRD." The same check has to be repeated for banks 1, 2 and 3 respectively. This can be done easily with a generate statement and a "for" loop as shown below.

```
genvar j;
generate
```

```
for (j=0; j<4; j++)
begin:loop
a_generate: assert property(@(posedge clk)
  (`s_active && bank_addr[1:0] == j)
  |-> not ##[1: tRRD] (`s_active &&
                (bank_addr[1:0] != j)));
c_generate: cover property(@(posedge clk)
  (`s_active && bank_addr[1:0] == j)
    |-> not ##[1: tRRD] (`s_active &&
                (bank_addr[1:0] != j)));
end
endgenerate
```

SDRAM_chk13: If "data_size" is 128, then check the mask operation.

The CPU-AHB bus can define the data size and write to the memory in 128 bits, 64 bits or 32 bits. The most commonly used data size is 32 bits. But when 128/64 bits are used, the mask bits are used to write the data in 32-bit chunks to the same address.

```
property p_xfer128;
@(posedge clk)
((size == 0) && ((dqm[0] == 0 && (`s_write
||`s_burst_write))) |->
  ##2 ($fell (dqm[1]) && addr == $past (addr, 2)
&&
  (`s_write ||`s_burst_write))
  ##1 $rose (dqm[1])
  ##1 ($fell (dqm[2]) && addr == $past (addr, 2)
&&   (`s_write ||`s_burst_write))
  ##1 $rose (dqm[2])
  ##1 ($fell (dqm[3]) && addr == $past (addr, 2)
 && (`s_write ||`s_burst_write))
  ##1 $rose (dqm[3]));
endproperty

a_xfer128: assert property(p_xfer128);
c_xfer128: cover property(p_xfer128);
```

Figure 5-21 shows the 128-bit data transfer. Data is written in 4 chunks of 32 bits to the same address location. Marker 1 shows the first 32 bits of data being written to address 0x0021 and marker 2 shows the fourth chunk of 32 bits of data being written to address 0x0021. Mask Bits dqm[3:0] are used to control the data that is being written to the memory.

- When the data transfer size is 32, all the bits of the vector dqm[3:0] are set to 0.
- If the data transfer size is 64, data is transferred in two chunks of 32 bits. When the first chunk of 32-bit data is transferred, dqm[1:0] is set to 0. When the second chunk of 32-bit data is transferred, dqm[3:2] is set to 0.
- For 128-bit transfers, when dqm[0] is set to 0, the first 32 bits of data is written to the memory and when dqm[1] is asserted, the second 32 bits of data is written to the memory. Similarly, when dqm[2] and dqm[3] are set to 0, the third and fourth chunks of 32 bits of data are written to the memory respectively.

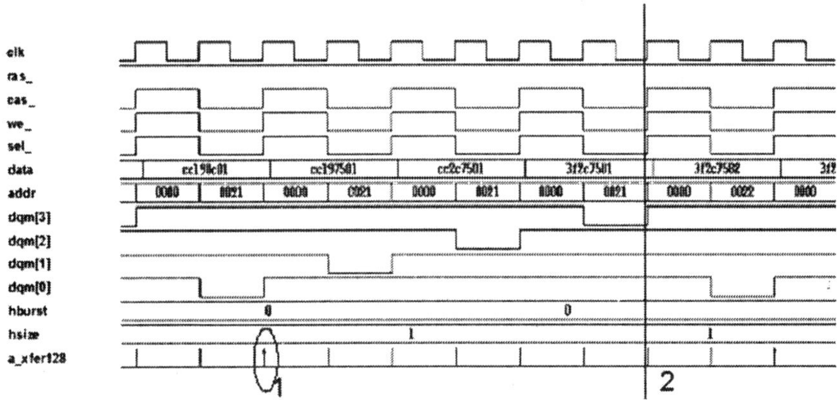

Figure 5-21. 128-bit data transfer

SDRAM_chk14: If "data_size" is 64, each read/write operation takes 2 cycles.

This property is similar to the previous one. The data is written in two chunks of 32 bits. When dqm[0] == 0 and dqm[1] == 0, the first chunk of 32 bits of data is written. When dqm[2] == 0 and dqm[3] == 0, the second chunk of 32 bits of data is written.

Figure 5-22 shows the 64-bit data transfer. The first 32 bits of data is written to address 0x0103 and mask signals 0 and 1 are set to 0 (marker 1).

The second chunk of data is written to the same address 0x0103 and mask signals 2 and 3 are set to 0 (marker 2).

```
property p_xfer64;
@(posedge clk) ((size == 1) && ((dqm[1:0] == 0
&&(`s_write ||`s_burst_write))) |->
##2 ($fell (dqm[2] &&  dqm[3]) && addr == $past
(addr, 2) && (`s_write ||`s_burst_write))
##1 $rose (dqm[3] && dqm[2]));
Endproperty

a_xfer64: assert property(p_xfer64);
c_xfer64: cover  property(p_xfer64);
```

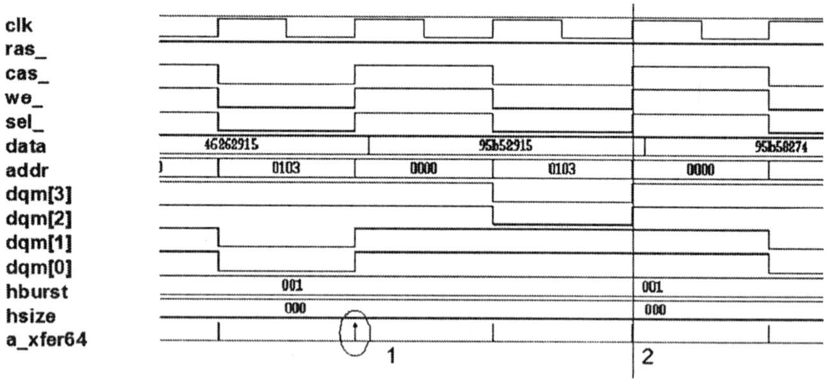

Figure 5-22. 64-bit data transfer

SDRAM_chk15: Read/write terminated by a burst terminate.

A burst terminate command is used to terminate a burst read/write command. So, if a burst terminate command is issued, the previous cycle must be a burst read/write operation.

```
property p_wr_rd_burstterminate;
@(posedge clk) (s_burstterminate) |->
$past ((`s_burst_write || `s_write || `s_read ||
`s_burst_read), 1);
endproperty
```

```
p_wr_rd_bursttterminate:
        assert property(p_wr_rd_bursttterminate);
c_wr_rd_bursttterminate:
        cover property(p_wr_rd_bursttterminate);
```

SDRAM_cover_chk1: Write terminated by a burst terminate.

There are some scenarios and properties that should be covered as part of the verification. For example, in the property (p_wr_rd_burstterminate), if a "burst terminate" command is issued, the previous command should be a "burst write" or a "burst read" command. But in the result of the check there is no classification on which specific command (read/write) was terminated by the "burst terminate" command, since all possible legal conditions have been combined with the logical OR operator. In order to obtain this kind of scenario information, the property is split and cover statements are written.

A separate property is written to check if the "burst write" was terminated using burst terminate. If this property is asserted, there might be failures because "burst terminate" command can be issued for terminating "burst read" also. Hence, for collecting coverage information on scenarios, there is no need to declare assert statements.

```
property p_wr_bursttterminate;
@(posedge clk)
(s_bursttterminate) |->
        $past ((`s_burst_write || `s_write), 1);
endproperty

c_wr_bursttterminate:
        cover property(p_wr_bursttterminate);
```

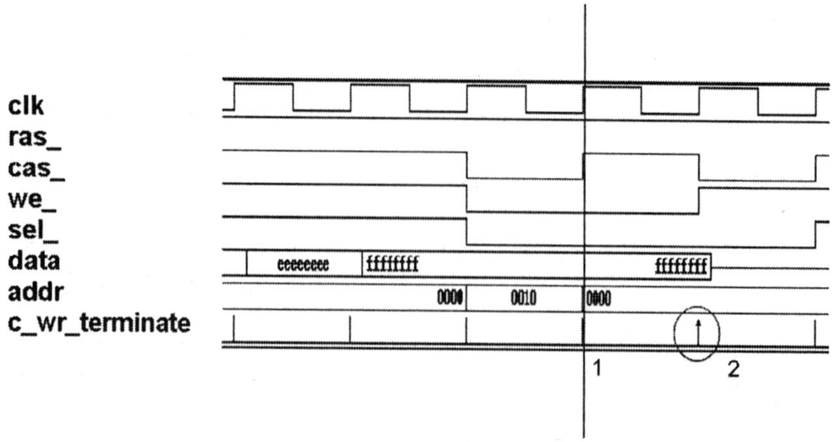

Figure 5-23. Burst write to Burst terminate command

Figure 5-23 shows a write command sampled at marker 1 and a burst terminate command sampled at marker 2. The cover statement is successful since the previous cycle of the burst terminate was a burst write command.

SDRAM_cover_chk2: Read terminated by a burst terminate.

The read command can be terminated by the "burst terminate" command similar to the previous check. This assertion checks that if in the current cycle a burst terminate command is issued than the previous cycles is a read/burst read.

Figure 5-24 shows a read command (marker 1) terminated by the "burst terminate" command (marker 2). But since there is "cas" latency to the read command, the data for the terminated read will be available two cycles later from when the read command was issued, as shown by marker 3. Until the data is available, no other command can be issued.

```
property p_rd_burstterminate;
        @(posedge clk) (s_burstterminate) |->
           $past  ((`s_burst_read   ||   `s_read),
        1);
endproperty

c_rd_burstterminate:
           cover property(p_rd_burstterminate);
```

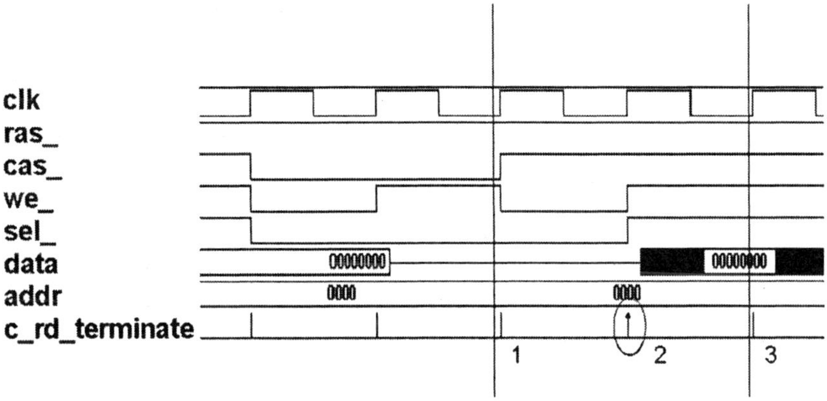

Figure 5-24. Read to Burst terminate

SDRAM_cover_chk3: Write terminated by a read.

The write command can be terminated by a read command. If a write command is in progress and a read command is issued, the write is immediately aborted. Once again, there is no need to assert this property. The write command can either be terminated by other ways or can be followed by any other command. Hence, asserting this property might produce unnecessary failures.

Figure 5-25 shows that a write command (marker 1) is being terminated by the read command (marker 2) and the read command is terminated by the burst terminate command. The burst size in this example is 4 and both the read/write operations are being terminated after just one write/read transfer.

```
property p_wr_rdterminate;
 @(posedge clk) (`s_write ||
  `s_burst_write) ##1 (`s_read || `s_burst_read);
endproperty

c_wr_rdterminate :
          cover property(p_wr_rdterminate);
```

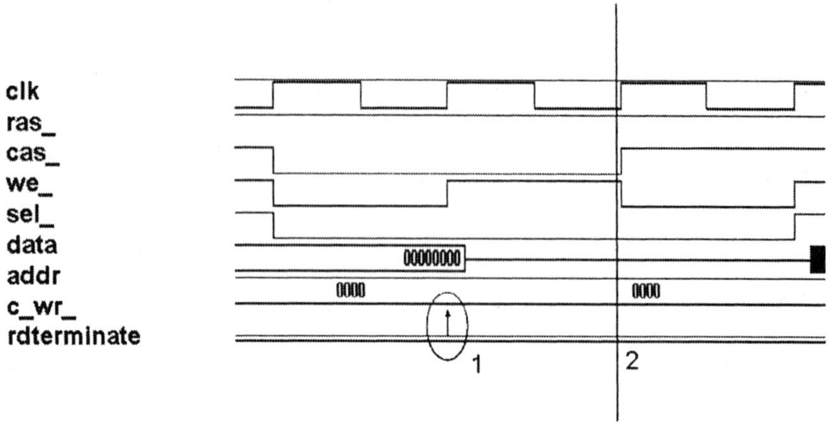

Figure 5-25. Write terminated by a Read command

5.3 SRAM/FLASH Verification

- SRAM (static RAM) is a type of memory that holds data without external refresh as long as it is powered.
- SRAMs are a lot faster than SDRAMs.
- SRAMs are expensive and take more space/area.

The verification of an SRAM/FLASH is very simple as there are no complex refresh mechanisms. When the "write" and "chip select" signals are asserted, the data is written in the memory starting from the location specified in the address bus. Similarly, when chip select is asserted, write enable is de-asserted and output enable is asserted, the data is read out of the memory. In the sample system, data cannot be written to the flash, since write protect signal is always asserted. The sample system uses these static memories.

SRAM : 256K X 16bit high speed Static RAM

FLASH: 128Mbit flash (16Mbytes)

The timing parameters of the SRAM used in the sample system are shown in Table 5-3. The timing parameters of the Flash used in the sample system are shown in Table 5-4.

Table 5-3. Timing parameters for SRAM

Parameter	Symbol	Min	Max
Write Cycle Time	tWC	1 cycle (10ns)	-
Write Pulse Width	tWP	2 cycle (20ns)	-
Read Cycle Time	tRC	1 cycle (10ns)	-
Chip Select to output	tCO	1 cycle (10ns)	-
Address Access time	tAA	1 cycle (10ns)	-

Table 5-4. Timing parameters for Flash memory

Parameter	Symbol	Min	Max
Read/Write Cycle time	tAVAV	15 cycles (150ns)	-
Chip select to Output Delay	tELQV	-	15 cycles (150ns)
Page Address Access time	tAPA	-	3 cycles (25ns)

5.3.1 SRAM/FLASH Assertions

SRAM_chk1: Write cycle time, tWC.

The SRAM write cycle time should be greater than the "tWC" mentioned in the specification. The write cycle time is the time in which the address is stable and in which the chip select and write enable signals are asserted.

To implement this assertion, the **$stable** system function and the implication operator are used. The **$stable** function makes sure that the value of the address in the current clock cycle is the same as the previous cycle.

```
property p_tWC;
@(posedge clk)
($fell (we_n) && !sel_n[2]) |=>
                  $stable(addr[22:0]);
endproperty
a_tWC: assert property(p_tWC);
c_tWC: cover property(p_tWC);
```

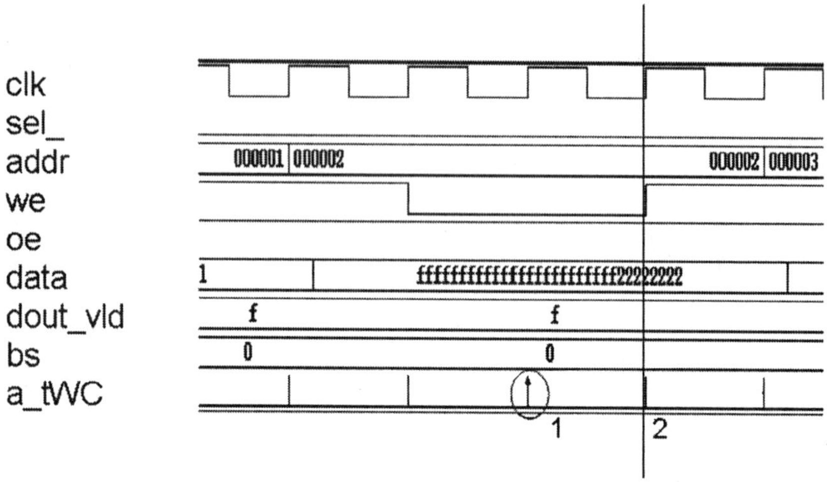

Figure 5-26. Write wycle time, tWC

Figure 5-26 shows that a write command is sampled at marker 1 and the assertion succeeds at marker 2, since the address is stable for at least one clock cycle from when the write was issued.

SRAM_chk2: Write enable pulse width, tWP.

This check verifies that the write pulse width is always greater than the minimum specified in the specification (2 cycles). Figure 5-27 shows that the falling edge of write (marker 1) and the rising edge of write (marker 2) are sampled 2 cycles apart.

```
property p_tWP;
@(posedge clk)
$fell (we_n) |->
        ##tWP $rose (we_n);
endproperty

a_tWP: assert property(p_tWP);
c_tWP: cover property(p_tWP);
```

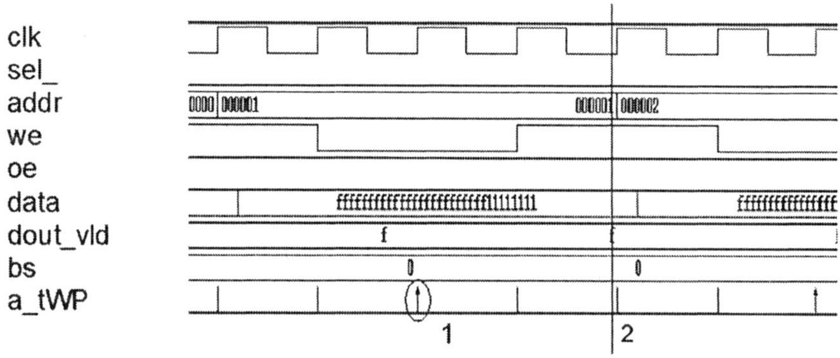

Figure 5-27. Write pulse width, tWP

SRAM_chk3: tRC - read cycle time.

This is similar to the write cycle time check. The read cycle time is the time in which the address is stable and in which the chip select and output enable are asserted. Figure 5-28 shows that chip select and output enable are asserted at marker 1. The address value is the same at both marker 1 and marker 2.

```
property p_tRC;
@(posedge clk)
(!sel_n[2] && we_n && !oe_n) |=>
                    ($stable (addr));
endproperty

a_tRC: assert property(p_tRC);
c_tRC: cover property(p_tRC);
```

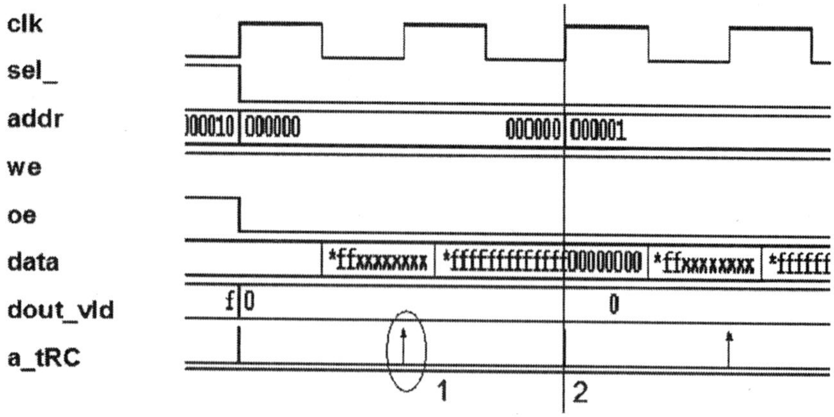

Figure 5-28. Read cycle time, tRC

SRAM_chk4: tCO - chip select to output data valid.

The parameter "tCO" is the minimum time that chip select has to be asserted before data becomes valid. Figure 5-29 shows that the chip select and output enable are asserted at marker 1. In the same clock cycle, the data value is "x." One cycle later, the data value is valid (marker 2).

```
property p_tCO;
@(posedge clk)
(!sel_n[2] && we_n && !oe_n &&
($isunknown (data))) |=>

          ($isunknown (data))==0;
endproperty

a_tCO: assert property(p_tCO);
c_tCO: cover property(p_tCO);
```

SRAM_chk5: tAA - Valid address to valid data.

The parameter "tAA" is the minimum time for which address has to be valid before data becomes valid. Figure 5-30 shows that a read command is sampled at marker 1. The address in this clock cycle should be stable in the next clock cycle and the data should be valid as shown by marker 2.

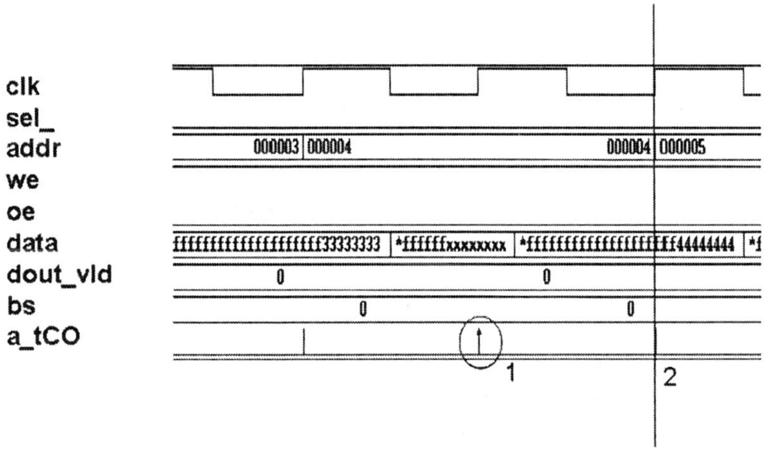

Figure 5-29. Chip Select to valid data, tCO

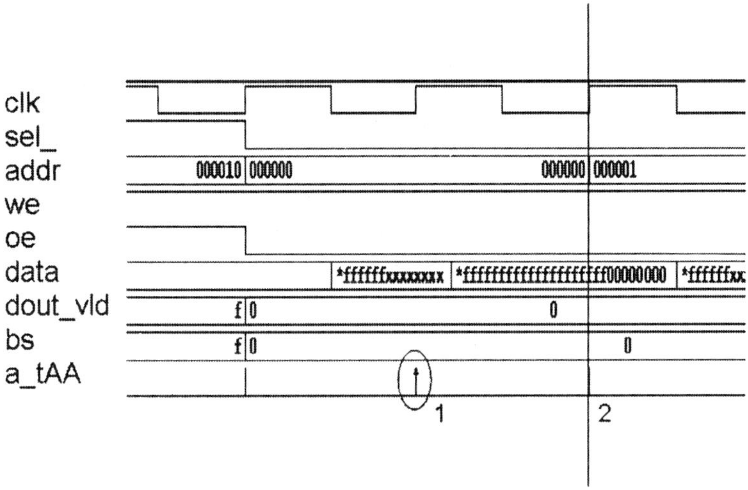

Figure 5-30. Valid address to Valid data, tAA

```
property p_tAA;
@(posedge clk) (!sel_n[2] && we_n && !oe_n) |=>
            ((addr == $past (addr,1))   ##0
            ($isunknown (data))==0);
endproperty
```

```
a_tAA: assert property(p_tAA);
c_tAA: cover property(p_tAA);
```

FLASH_Chk1: Flash is write protected.

The flash memory used in the sample system is write protected. It is necessary to make sure that the write protect signal (wp_) is always asserted when the chip select for flash is enabled.

```
property p_write_protect;
 @(posedge clk)
(!sel_n[3]) |->
                wp_n == 0;
endproperty

a_write_protect:
       assert property (p_write_protect);
c_write_protect:
       cover property (p_write_protect);
```

FLASH_chk2: Complete read cycle time (tAVAV).

The minimum read cycle time as mentioned in the Table 5-4 is tAVAV(15 clock cycles). In the sample system used, the read cycle time cannot be more than 900 clock cycles. Hence, two checks are written to verify both the minimum and maximum timing requirements.

```
property p_tAVAV_not;
  @(posedge clk)
(!sel_n[3] && $fall(oe_n)) |->
                    not ##[0:15] $rose (oe_n);
endproperty

property p_tAVAV;
  @(posedge clk)
  (!sel_n[3] && $fell (oe_n))   |->
                  ##[16:900] $rose (oe_n);
endproperty

a_tAVAV: assert property(p_tAVAV);
a_tAVAV_not: assert property(p_tAVAV_not);

c_tAVAV: cover property(p_tAVAV);
```

```
c_tAVAV_ not: cover property(p_tAVAV_not);
```

FLASH_chk3: CS/ADDR to valid data is tELQV.

The minimum time that the chip select and address should be stable before data is valid is specified by "tELQV" (15 clock cycles). Figure 5-31 shows that the signal "sel" is asserted at marker 1. After 15 cycles, the first data is valid as denoted by marker 2.

```
property p_tELQV;
@(posedge clk)
(!oe_n && $fell (sel_n[3])) |->
##14 $isunknown(data) ##1 ($isunknown (data)==0);
endproperty

a_tELQV: assert property(p_tELQV);
c_tELQV: cover property(p_tELQV);
```

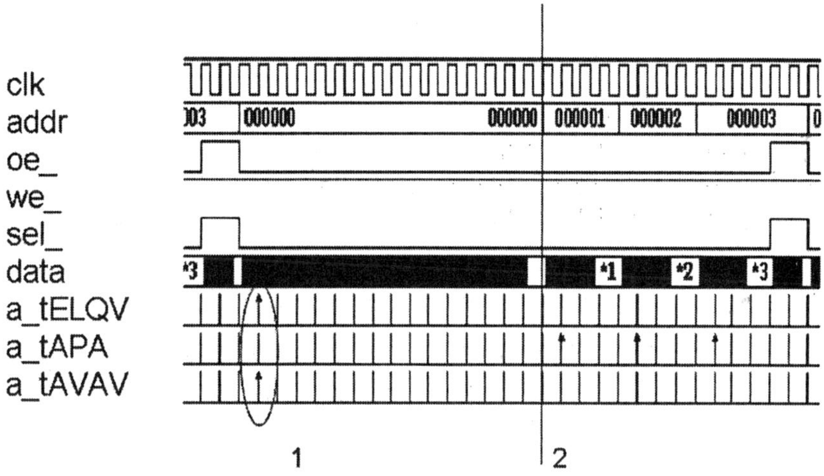

Figure 5-31. Flash waveform for tELQV, tAPA, tAVAV

FLASH_chk4: ADDR to valid data, tAPA.

The parameter tAPA (3 cycles) is the minimum time that the address is required to be stable before data is valid. This is true for only the subsequent reads of the burst read and not the first read in a burst.

Figure 5-32 shows a burst read command. In a burst read, when the address changes, a new data should be read within tAPA. The change in address from "000000" to "000001" is sampled at marker 1. At this point, the data is unknown and 3 clock cycles later a valid data is sampled, as shown by marker 2.

```
sequence s_data_trans;
(!sel_n[3] && !oe_n && ($stable (addr)==0) &&
$stable (oe_n)) ##0 $isunknown (data)
##3 $isunknown (data)==0;
endsequence

property p_tAPA;
@(posedge clk)
s_data_trans |->
          $stable(addr);
endproperty

a_tAPA: assert property(p_tAPA);
c_tAPA: cover property(p_tAPA);
```

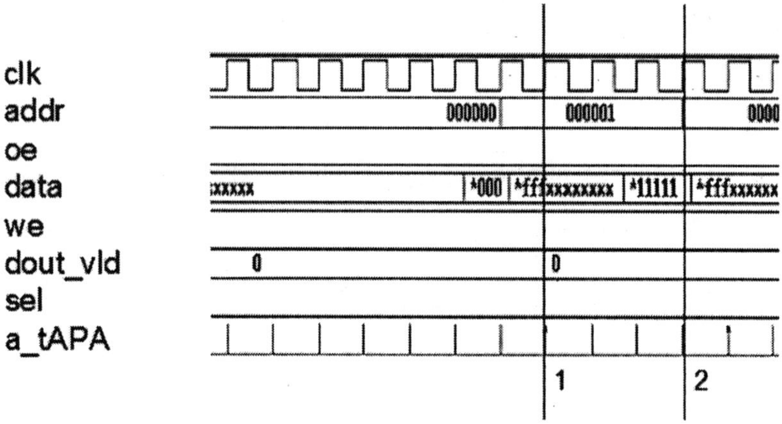

Figure 5-32. Flash waveform for tAPA

5.4 DDR-SDRAM Verification

Double Data Rate Synchronous Dynamic Random Access Memory (DDR-SDRAM) is a type of memory that is similar to Synchronous DRAM but has a higher bandwidth. Data is written and read at both the rising and falling edge of the clock, doubling the speed. The operations are similar to that of the SDRAM. The DDR-SDRAM used in the sample system has the following configuration.

DDR-SDRAM: 4Mword x 16bit x 4bank

The read and burst read operation in the DDR-SDRAM is the same as that of the SDRAM. The burst read command is issued by asserting "sel_" and "cas_" while holding "ras_" and "we_" high. The address inputs determine the starting address of the burst. The first data is available after the "cas" latency after the read command (which is 2 cycles, based on the DDR-SDRAM specifications), and the subsequent data are presented on the rising and falling edges of the signal "dqs" (data strobe).

The burst write command is issued by asserting "sel_," "cas_," "we_" and de-asserting "ras_" on the rising edge of the clock (clk). There is a latency of 1 clock cycle for the signal "dqs" to arrive. There is no latency relative to the signal "dqs" for a write command.

5.4.1 DDR-SDRAM Assertions

DDR_Chk1: Burst Read operation for DDR memories.

In the DDR memory there are multiple clocks. Data transfer and read are done on clock "clk2x" which samples data on both edges. Most of the checks written for SDRAM can be reused for a DDR-SDRAM. New checks have to be written wherever the control signals are crossing clock domains.

The keyword **matched** is used to synchronize signals across multiple clock domains in SVA. In this assertion, the signals cas_, ras_, we_ and sel_ are being generated by clock "clk." The data is read at the negative edge of clock "clk2x." In this case, we have to use the **matched** construct to synchronize the read sequence from one clock domain to another.

```
sequence s_read;
    @(posedge clk)
    (ras_n && !sel_n[0] && we_n && !cas_n);
```

```
endsequence

property p_read;
@(negedge clk2x) s_read.matched |->
##3 ($isunknown (data))
##1 ($isunknown (data) == 0);
endproperty

a_read: assert property(p_read);
c_read: cover property(p_read);
```

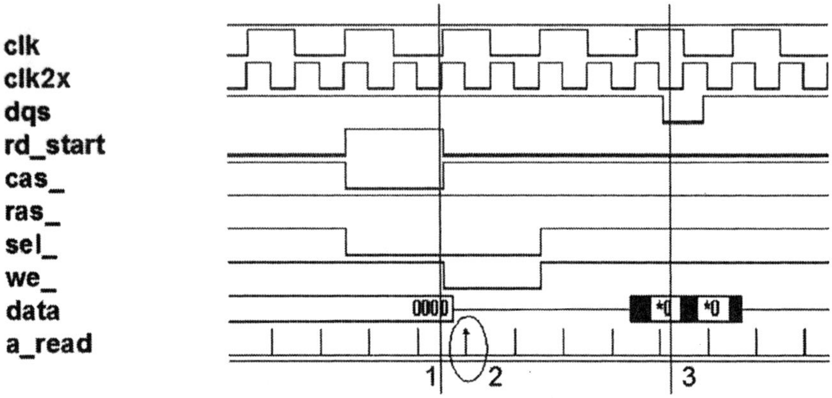

Figure 5-33. DDR-SDRAM Burst read operation

Figure 5-33 shows that a read command is sampled at marker 1 (s_read) based on the clock "clk." The matched value of this sequence is sampled in the next nearest negative edge of clock "clk2x," as shown by marker 2. Data is then read out with a CAS latency of 4 clock cycles (clk2x) denoted by marker 3. A valid data is read on both edges of the signal "dqs" and the signal "dqs" is generated based on clock (clk2x). Hence, the negative edge of the clock (clk2x) is used to sample the data.

DDR_Chk2: Burst write operation on DDR memories.

```
sequence s_write;
 @(posedge clk)
   (ras_n && !sel_n[0] && !we_n && !cas_n);
endsequence
```

```
property p_write;
  @(posedge clk2x) s_write.matched
      |->  ##1 ($isunknown(data) == 0)
           ##1 ($isunknown(data) == 0);
endproperty

a_write: assert property(p_write);
c_write: cover property(p_write);
```

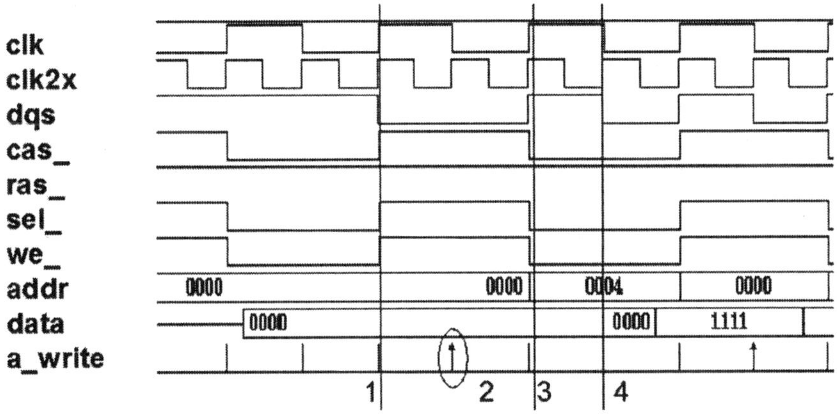

Figure 5-34. DDR-SDRAM Burst write operation

Figure 5-34 shows a write command at marker 1 (based on clock (clk)). The write command is synchronized to the positive edge of clk2x at marker 2. The positive edge of the clock (clk2x) is used for sampling in a write command because the signal "dqs" is generated based on the clock (clk2x). The assertion is successful at marker 2 since data is being written into the memory on both edges of the signal "dqs" (data strobe) as shown by marker 3 and marker 4.

5.5 Summary on SVA for Memories

- Assertions can be used effectively to verify the timing requirements of memory devices.
- All timing information relevant to the memory device should be parameterized. This way, the assertions developed for a particular type of memory can be reused with similar memory device from any other vendor.

- The assertions written for memories provide information on specific **scenario coverage**. For example, was a write terminated by a read/burst terminate, was a read terminated by a burst terminate, was a back to back write performed, did a write command follow a read command, did the tests cover different data widths - 128/64/32 bits, etc. This helps increase the verification confidence and also provides a measure for verification completeness.

Chapter 6

SVA FOR PROTOCOL INTERFACE

SVA checkers for a sample PCI system

Compliance testing has become one of the major challenges in SOC designs. It is very common for designs to support certain standard protocols. For example, graphics applications might support a standard bus interface such as PCI/PCIX, USB or IEEE 1394 Firewire. These bus interfaces help the designs achieve higher bandwidth of data transmission and also provide a standard method to connect multiple devices. Bus protocols are complex and every device sitting on the bus should be compliant with a list of rules specific to that protocol.

The verification environment built for testing these standard protocol interfaces are often re-usable since the same set of rules applies to any device that supports the specific interface. Verification engineers often develop bus interface models (BIM) of the devices that support a specific interface. The BIM need not replicate the detailed internal functionality of the device. It just has to support the basic handshaking process that is compliant with the specific interface. This helps the verification engineer to create a sample system with the BIM and the Design Under Test (DUT). Tests can be written to create transactions between the BIM and the DUT. While running these tests, specific monitors are written to make sure that the DUT is being absolutely compliant with the standard protocol. Most verification environments create logs of the transactions as seen by the bus. SVA can be used very effectively to create these bus protocol monitors. In this chapter, a sample PCI system is used to demonstrate how SVA checkers are created for a PCI compliant device.

6.1 PCI – A Brief Introduction

The PCI local bus is a high performance, 32-bit or 64-bit bus with multiplexed address and data lines. The bus is intended for use as an interconnect mechanism between highly integrated peripheral controller components, peripheral add-in boards and processor memory systems. A sample PCI compliant device is shown in Figure 6-1.

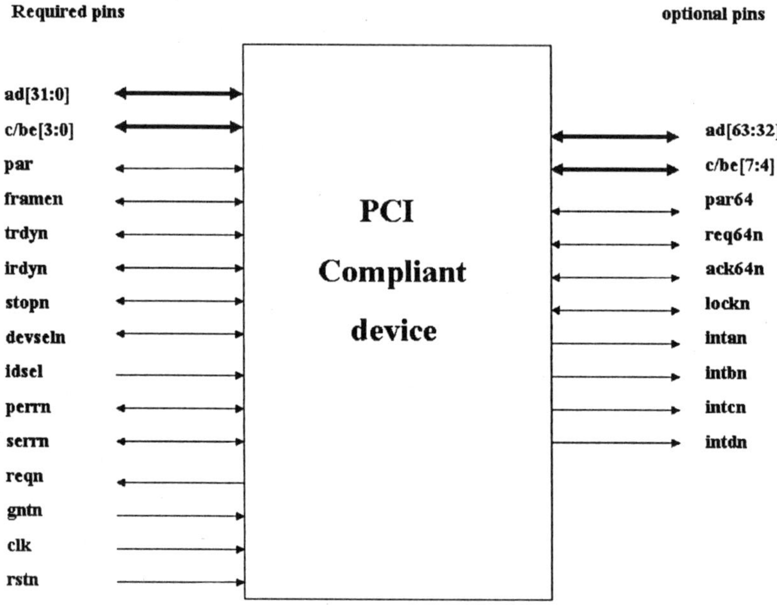

Figure 6-1. PCI compliant device

A brief description of each pin is listed below.

ad[31:0] – the address bus, this has the information on the location to which data is to be transferred or the location from which data should be obtained. This also acts as the data bus.

c/be[3:0] – the command bus, contains one of the twelve commands shown in Table 6-1. It also acts as the byte enable bus that defines which bytes in the data bus are to be transferred.

par – a parity bit, even number of 1's should appear on the ad, c/be and par bits. The required value of the par bit is driven by the device one clock cycle after the device drives the "ad" bus.

framen – frame signal, this is asserted by the master that wants to perform a data transaction. When the frame is asserted, the master also indicates the nature of the transaction by setting the appropriate command on the "c/be" bus. The frame signal is de-asserted when the master is ready to complete the final data transfer.

Table 6-1. PCI Bus commands

C/BE[3:0]	Command type
0000	Interrupt Acknowledge
0001	Special cycle
0010	I/O Read
0011	I/O write
0100	Reserved
0101	Reserved
0110	Memory Read
0111	Memory Write
1000	Reserved
1001	Reserved
1010	Configuration Read
1011	Configuration Write
1100	Memory Read Multiple
1101	Dual Address cycle
1110	Memory Read Line
1111	Memory Write and Invalidate

trdyn – target ready signal, this is asserted by the target device that is currently addressed by the master. By asserting this signal the target device lets the master know that it is ready for a data transaction.

irdyn – master ready signal, this is asserted by the master that wants to perform a data transaction.

stopn – stop signal, this is asserted by the target device if it wants to terminate the current transaction. If the target asserts the stop signal without performing any data phases, it is called a retry. If the target asserts the stop signal after performing one or more data phases, it is called a disconnect.

devseln – device select signal, this is asserted by the target device if it is selected. The target ready signal is asserted only after asserting this signal.

idsel – initialization device select signal, is used as a chip select during PCI configuration read and write transactions.

perrn – parity error signal, asserted one clock after a parity error is identified either by the master or a target.

serrn – system error signal, it is an output of both master and target devices. This is asserted only when something fatal occurs.

reqn – request signal, this is used by the master device to request the use of the PCI bus.

gntn – grant signal, this indicates that the PCI device has got the permission to use the PCI bus.

ad[63:32] – upper 32 bits of the data bus, used for 64-bit transactions.

c/be[7:4] – acts as the byte enable bus that defines which of the bytes in the upper data bus is to be transferred in a 64-bit transaction.

par64 – a parity bit, even number of 1's should appear on the ad[63:32], c/be[7:4] and par64 bits. The required value of the par64 bit is driven by the device one clock cycle after the device drives the "ad" bus.

req64n – request signal, this is used by the master device to request the use of the PCI bus for a 64-bit transaction.

ack64n – acknowledge pin, PCI target device acknowledges the 64-bit transaction requested by the master device.

6.1.1 A sample PCI Read transaction

A read transaction is initiated by the master device. The master asserts the "framen" signal and drives an address onto the "ad" bus. It also places a read command on the "c/be" bus. The target device decodes the address and identifies itself. Once it identifies itself, it asserts the "devseln" signal. The master device continues asserting the "framen" signal, but stops driving the address bus. It asserts the signal "irdyn" and also places the byte enable command on the "c/be" bus. In response to this, the target device places the first data on the data bus (ad) and also asserts the signal "trdyn" to acknowledge that the data on the bus is valid. In a multiple data transaction, it is the responsibility of the addressed target to increment the initial address to point to the subsequent data locations.

During a transaction if the target device is not ready to place the next data on the bus, it creates a wait state by de-asserting the signal "trdyn." The signal "devseln" will stay asserted and the data placed on the bus in the previous transaction will stay. The master will read the data only if both "trdyn" and "irdyn" are asserted. When the target is ready to transmit again it will assert the signal "trdyn." The master device indicates that the next data read will be the last one in the current transaction by de-asserting the signal "framen." Once the last data is read, the master de-asserts the signal "irdyn" and the target device de-asserts the "trdyn" and the "devseln" signals respectively. If both the signals "irdyn" and "framen" are de-asserted, the bus is said to be in an idle state. A waveform for a sample PCI read transaction is shown in Figure 6-2.

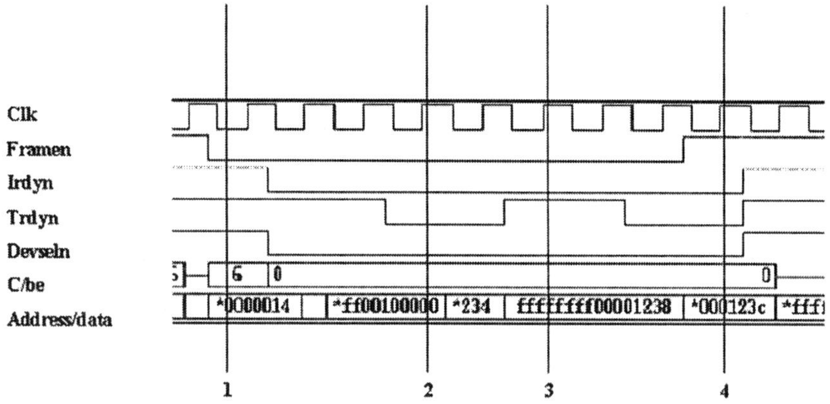

Figure 6-2. Sample PCI read transaction

Marker 1 shows the point where the master issues the read command (0110) on the "c/be" bus. On the same cycle, the address bus also carries the address for the target device. Marker 2 shows the point when the master reads a valid data. Marker 3 shows that the target device de-asserts the "trdy" signal indicating that it is not ready for the read transaction. Marker 4 indicates the last data phase since the signal "framen" is de-asserted. In the next clock cycle, the signals "irdyn," "trdyn" and "devseln" are all de-asserted, indicating the completion of the transaction.

6.1.2 A sample PCI Write transaction

The master device initiates a write transaction. It asserts the signal "framen" and drives an address onto the "ad" bus. It also places the write command on the "c/be" bus. The target device identifies itself and asserts the signals "devseln" and "trdyn." The master continues to assert the "framen" signal. The master places the data on the "ad" bus and also asserts the signal "irdy" to let the target know that the data on the bus is valid. The master also issues the command byte that identifies which bytes are to be written on the same clock cycle. If the master is not ready to place the next data on the bus, it can create a wait state by de-asserting the signal "irdyn." The master will drive the same data from the previous cycle during the wait state. The master will de-assert the "framen" signal just before the last data is ready to be written. Once all the data is written, the master de-asserts the "irdyn" signal and then the target de-asserts the signals "trdyn" and "devseln." A waveform for a sample PCI write transaction is shown in Figure 6-3.

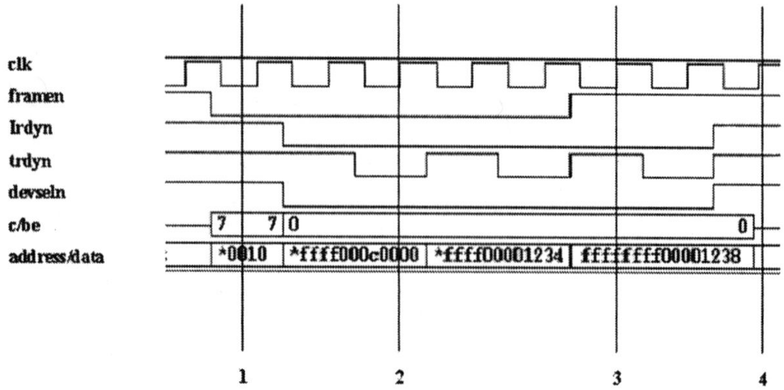

Figure 6-3. Sample PCI write transaction

Marker 1 shows the point where the master issues the write command (0111) on the "c/be" bus. On the same cycle, the address bus also carries the address for the target device. Marker 2 shows the point when the master writes a valid data. Marker 3 shows that the target device de-asserts the "trdy" signal indicating that it is not ready for the write transaction. In the same clock cycle, the master device de-asserts the signal "framen" indicating that this is the last data phase. Marker 4 shows that the signals "irdyn," "trdyn" and "devseln" are all de-asserted, indicating the completion of the write transaction.

6.2 A sample PCI System

A sample PCI system used for illustration purpose is shown in Figure 6-4. The figure shows that there are 2 PCI master devices and 2 PCI target devices. A user could be designing a device that is expected to act as a PCI master or a PCI target or both. One could use bus interface models for the other three devices in the sample system to verify the DUT. There are three specific scenarios for which SVA checkers could be written as part of the verification plan. These three scenarios are discussed in the upcoming sections.

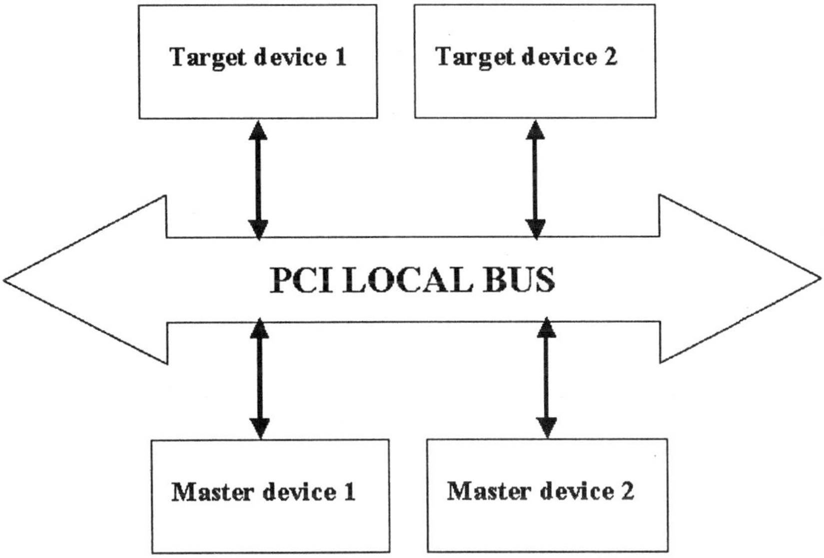

Figure 6-4. Sample PCI system

6.3 Scenario 1 – Master DUT Device

In this section, we assume that the design under test is a PCI master. Based on the PCI local bus specification, the PCI master has to follow certain protocol to be fully compliant. It is very common to write monitors as part of the verification environment. These monitors make sure that the DUT is not violating any of the protocol specifications.

The monitors can also produce detailed log files of all master transactions for post-processing purpose. SVA can be used to define a generic set of checkers that can be attached to any PCI master device. Since PCI is a standard protocol, the checkers developed should be written in such a way that it can be re-used with any PCI compliant master device. Figure 6-5 shows a sample configuration of the system.

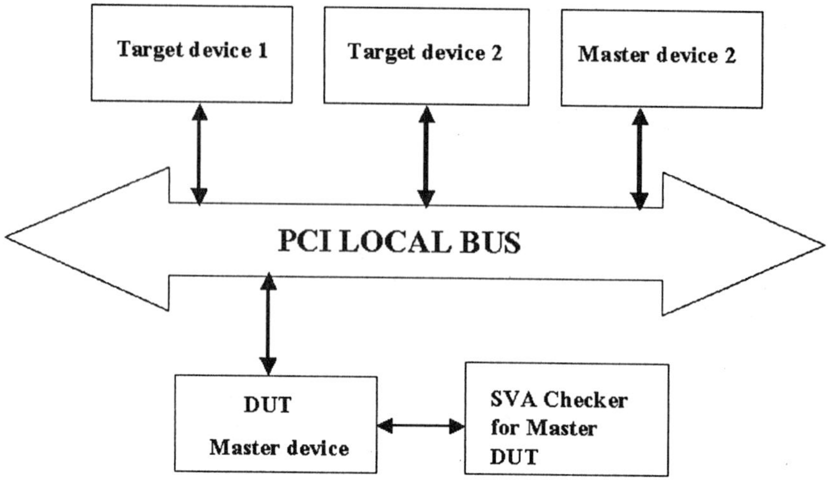

Figure 6-5. Sample configuration for PCI Master device as the DUT

6.3.1 PCI Master assertions

In this section, we show a few sample SVA checkers that can be written to verify the PCI master functionality. Some of the commonly used design conditions are defined as follows to enable re-use.

```
`define s_IO_READ
   ($fell (framen) && (cxben[3:0] == 4'b0010))
`define s_IO_WRITE
   ($fell (framen) && (cxben[3:0] == 4'b0011))
`define s_MEM_READ
   ($fell (framen) && (cxben[3:0] == 4'b0110))
`define s_MEM_WRITE
   ($fell (framen) && (cxben[3:0] == 4'b0111))
`define s_CONFIG_READ
   ($fell (framen) && (cxben[3:0] == 4'b1010))
`define s_CONFIG_WRITE
   ($fell (framen) && (cxben[3:0] == 4'b1011))
`define s_DUAL_ADDR_CYCLE
   ($fell (framen) && (cxben[3:0] == 4'b1101))
`define s_MEM_READ_LINE
   ($fell (framen) && (cxben[3:0] == 4'b1110))
`define s_MEM_WRITE_INV
   ($fell (framen) && (cxben[3:0] == 4'b1111))
```

```
`define s_BUS_IDLE
    (framen && irdyn)
```

Master_chk1: On a given clock cycle, "framen" cannot be de-asserted unless "irdyn" stays asserted on the same clock cycle.

```
property p_mchk1;
@(posedge clk)
    $rose (framen) |-> (irdyn == 0);
endproperty

a_mchk1: assert property(p_mchk1);
c_mchk1: cover property(p_mchk1);
```

The master device asserts the signal "framen" during the last data phase. Hence, the signal "irdyn" should stay asserted at this point. If not, this is a violation. Figure 6-6 shows a sample waveform of this check in a simulation. Markers 1, 2, 3, 4 and 5 show instances where there is a rising edge on the "framen" signal and in all those clock edges, the signal "irdyn" was always asserted. Hence, the checker succeeds.

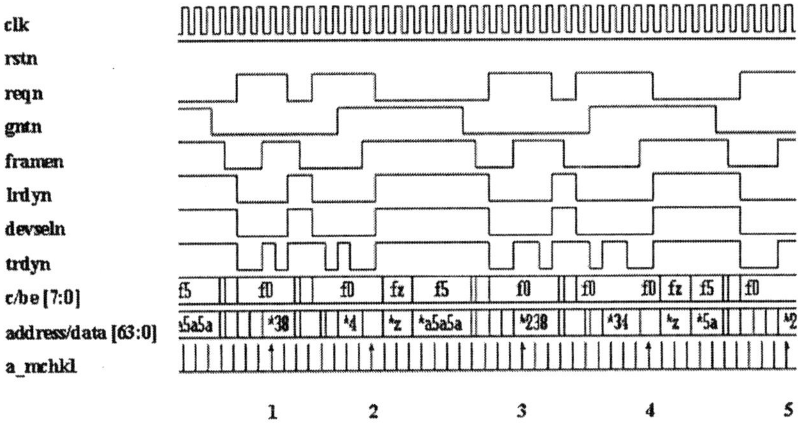

Figure 6-6. PCI Master check1

Master_chk2: Once "framen" is de-asserted, it cannot be asserted during the same transaction.

```
property p_mchk2;
@(posedge clk) $rose (framen) |->
      framen[*1:8] ##0 $rose (irdyn && trdyn);
endproperty

a_mchk2: assert property(p_mchk2);
c_mchk2: cover property(p_mchk2);
```

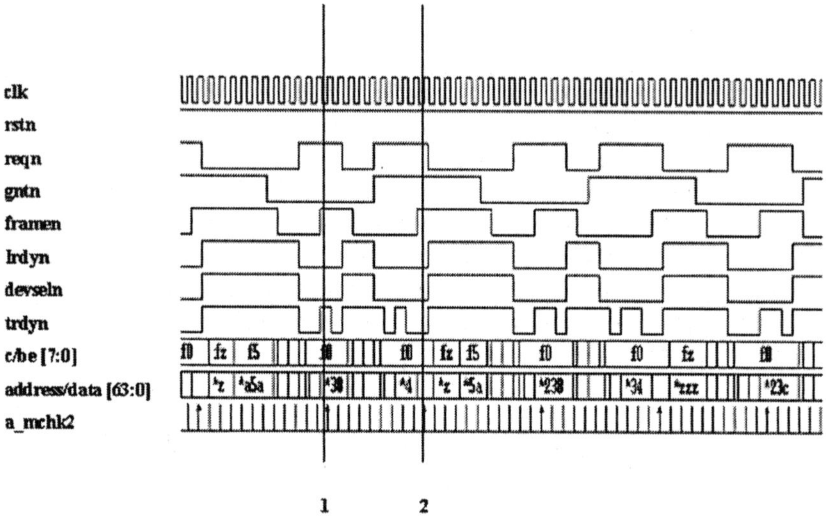

Figure 6-7. PCI Master check2

Once the signal "framen" is de-asserted, the master device has only one more data phase left. But it can take more than just one cycle to complete the last data phase. For example, if the target is not ready to accept the data, then the master waits to finish the last data phase. Before the master completes the last data phase, the frame cannot be asserted again. In other words, the signals "irdyn" and "trdyn" have to be de-asserted first before asserting "framen" again. Figure 6-7 shows a sample waveform of this check in a simulation.

Marker 1 shows a success of the assertion. At this point, there is a rising edge on the signal "frame" and hence the check becomes active. Note that in the same clock cycle, the signal "trdyn" is de-asserted indicating that the target is not ready to accept data. In the next clock cycle, both the signals "trdyn" and "irdyn" are asserted and hence the last data phase is complete.

One clock cycle, later the signals "irdyn" and "trdyn" are de-asserted. Marker 2 also shows a success but in this case, when the rising edge of the frame occurs, both the signals "irdyn" and "trdyn" are asserted and hence the last data phase is completed. In the next clock cycle, the signals "irdyn" and "trdyn" are de-asserted.

Master_chk3: Once "irdyn" is asserted, the master cannot change "irdyn" or "framen" until the current data phase begins.

```
property p_mchk3;
@(posedge clk)
$fell (irdyn) ##[0:5]
!(devseln) ##0 stopn    |->
      (!irdyn ) [*0:16] ##0 !trdyn;
endproperty

a_mchk3: assert property(p_mchk3);
c_mchk3: cover property(p_mchk3);
```

Once a master asserts the signal "irdyn," it is expected that a valid data phase begin within 16 clock cycles assuming there are no stop conditions issued by the target device. The data phase begins when the target device asserts the signal "trdyn." From the point when signal "irdyn" is asserted, assuming there are no stop conditions, the signal "irdyn" should be kept asserted until the signal "trdyn" is asserted by the target device.

Figure 6-8 shows a sample waveform of this check in a simulation. Marker 1 shows a success of the checker. The signal "irdy" and "devseln" are asserted at this point. One cycle later "trdyn" is asserted and hence the checker succeeds.

Marker 2 shows a condition wherein both signals "irdyn" and "devseln" are asserted and 2 clock cycles later, the signal "trdyn" is asserted. The signal "irdyn" stays asserted until the arrival of the "trdyn" signal and hence the checker succeeds.

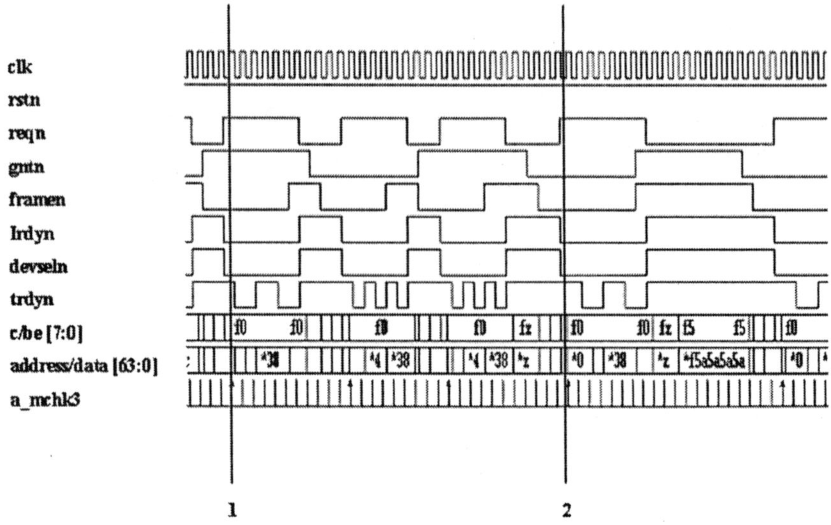

Figure 6-8. PCI Master check3

Master_chk4: The master is required to assert "irdyn" within 8 cycles from when the "framen" is asserted.

An **intersect** construct is used to control the length of the entire property. If the consequent of the property does not succeed within 1 to 8 clock cycles, the assertion will fail.

```
property p_mchk4;
@(posedge clk)
$fell (framen) |->
  1[*1:8] intersect
  ($fell (framen) ##[1:$] $fell(irdyn));
endproperty

a_mchk4: assert property(p_mchk4);
c_mchk4: cover property(p_mchk4);
```

Master_chk5: Normal Termination, once "framen" is de-asserted, the last data phase is completed within 8 clock cycles.

```
property p_mchk5;
@(posedge clk)
$rose (framen) |->
```

```
    (##[1:8] ($rose (irdyn && trdyn && devseln)));
endproperty

a_mchk5: assert property(p_mchk5);
c_mchk5: cover property(p_mchk5);
```

Master_chk6: Master Abort, "devseln" should be asserted within 5 cycles of "framen" being asserted. If "devseln" is not asserted within 5 cycles, then the "framen" should be de-asserted and one cycle later "irdyn" should be de-asserted.

```
sequence s_mchk6;
@(posedge clk)
  $fell (framen) ##1 (devseln)[*5] ##0 framen;
endsequence

property p_mchk6;
@(posedge clk)
  s_mchk6.ended |-> ##1 $rose (irdyn);
endproperty

a_mchk6: assert property(p_mchk6);
c_mchk6: cover property(p_mchk6);
```

Figure 6-9 shows a sample waveform of this check in a simulation. Marker 1 shows the clock edge in which the signal "framen" is detected as asserted. If the signal "devseln" does not arrive in the next 5 clock cycles, then the master should abort this transaction. Marker 2 shows the clock edge on which "devseln" failed to arrive and Marker 3 shows the next clock edge wherein the master device de-asserts the signal "irdyn." Since the property starts when the sequence s_mchk6 ends successfully, marker 2 is the point where the success is shown.

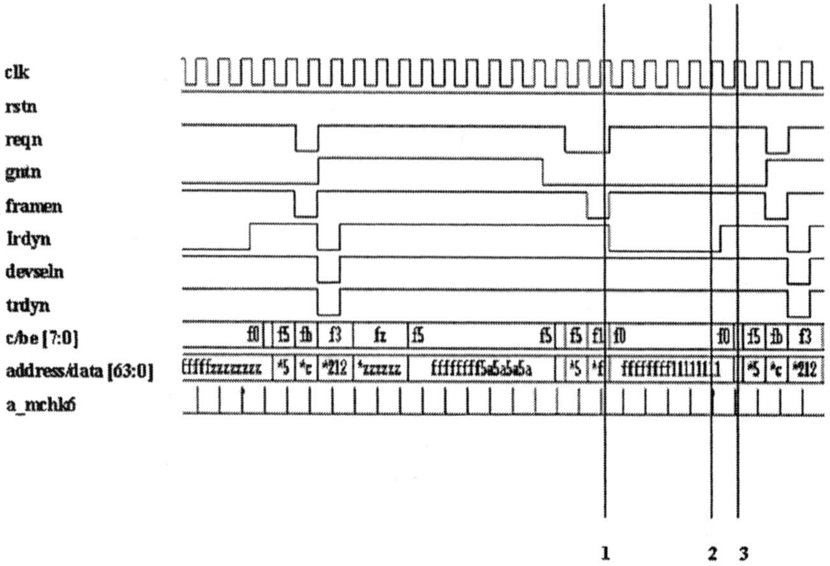

Figure 6-9. PCI Master check6

Master_chk7: When master is aborted by a target either by retry or disconnect, the master must de-assert its request before repeating the transaction. The request should be de-asserted on the clock cycle when the bus goes to the idle state and one clock cycle before or after the idle state.

```
sequence s_mchk7_before;
@(posedge clk)
    (!devseln && $fell (stopn) && trdyn)
    ##1 reqn ##1 `s_BUS_IDLE;
endsequence

sequence s_mchk7_after;
@(posedge clk)
    (!devseln && $fell (stopn) && trdyn)
    ##1 !reqn ##1 `s_BUS_IDLE;
endsequence

property p_mchk7_before;
@(posedge clk)
    s_mchk7_before.ended |->
                              reqn;
```

```
endproperty

property p_mchk7_after;
@(posedge clk)
   s_mchk7_after.ended |->
                        reqn [*2];
endproperty

a_mchk7_before: assert property(p_mchk7_before);
a_mchk7_after: assert property(p_mchk7_after);

c_mchk7_before: cover property(p_mchk7_before);
c_mchk7_after: cover property(p_mchk7_after);
```

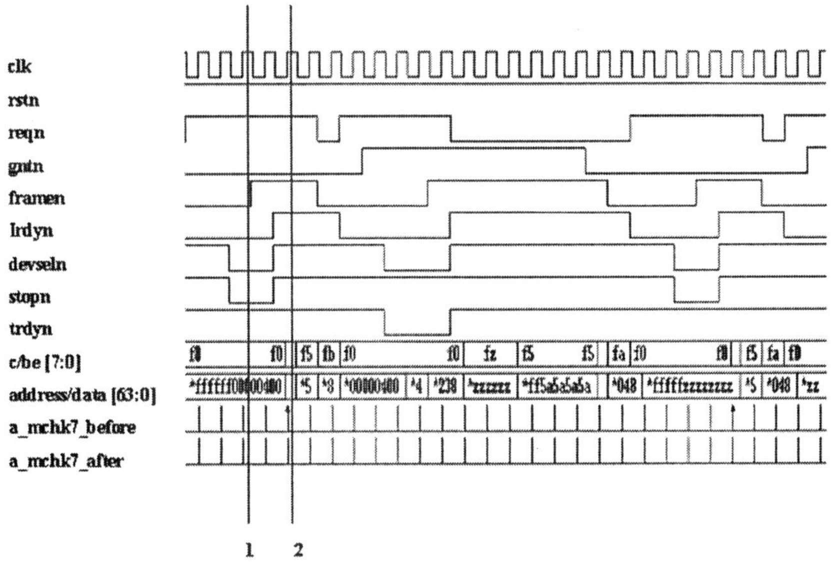

Figure 6-10. PCI Master check7

This check needs two separate properties. The main requirement is that, if the target device issues a stop condition, the master will have the "reqn" signal de-asserted before requesting the bus again. The master device could have de-asserted the signal "reqn" before the stop condition actually arrived. In this case, it is verified that the signal "reqn" is de-asserted during the clock cycle when the bus is idle and also that the signal "reqn" was de-asserted in the previous clock cycle (p_mchk7_before).

If the master device did not de-assert the signal "reqn" before the arrival of the stop condition, then it is verified that the signal "reqn" is de-asserted during the bus idle cycle and also the next clock cycle (p_mchk7_after). Note that the bus becomes idle 2 cycles after the target device asserts the signal "stopn." Figure 6-10 shows a sample waveform of this check in a simulation.

Marker 1 shows the clock edge when the target device asserts the "stopn" signal. Marker 2 shows the point when the bus becomes idle. Note that the signal "reqn" is de-asserted at this clock cycle and also in the previous clock cycle. Hence, the checker a_mchk7_before succeeds.

Master_chk8: When the target device terminates a transaction with a retry command, the master must repeat the same transaction until it is completed.

```
sequence s_mchk8a(temp1);
@(posedge clk)
(((!gntn || $rose (gntn))
&& $fell framen)),temp1=cxben[3:0])
##[1:2] $fell(irdyn) ##[0:5] $fell(stopn)
&& $fell (devseln) && trdyn;
endsequence

sequence s_mchk8b(temp2);
@(posedge clk)
$fell (reqn) ##[0:100] !gntn
##[0:5] $fell (framen)
##0 ((cxben[3:0] == temp2));
endsequence

property p_mchk8;
int temp;
@(posedge clk)
s_mchk8a(temp) |->
     ##[2:20] s_mchk8b(temp);
endproperty

a_mchk8: assert property(p_mchk8);
c_mchk8: cover property(p_mchk8);
```

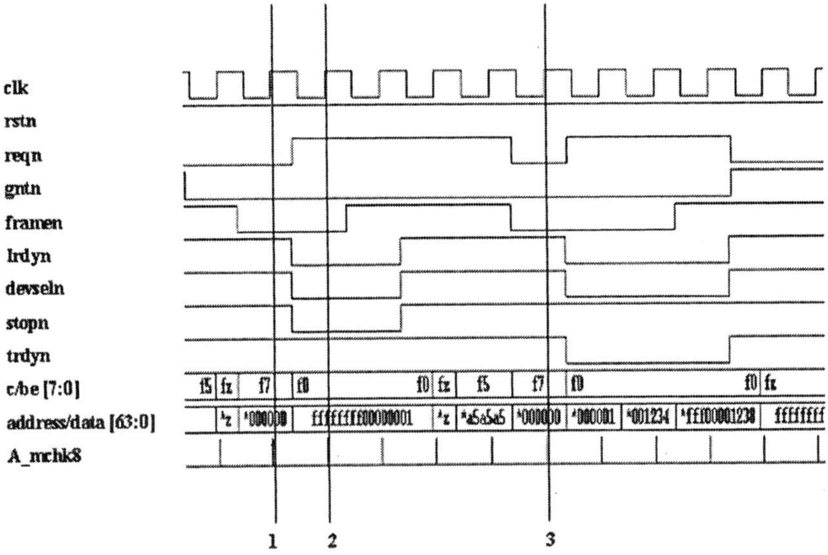

Figure 6-11. PCI Master check8

Two separate sequences are written to check this property. The first sequence starts at the point when the master device asserts the "framen" signal. When the master asserts the frame, it also issues the command. A temporary variable called "temp1" is used to store the command that was issued by the master. The variable is updated upon a successful match on a falling edge of the signal "framen." In the next few cycles, if the target device terminates the transaction by asserting the signal "stopn," then the sequence s_mchk8a will match. The property p_mchk8 has the sequence s_mchk8a as the antecedent. If the antecedent is true, then we wait for the next command to be issued by the master. If and when the master issues a new command, we compare the command value stored in the local variable "temp" to the actual command issued by the master on the bus, to verify that both the commands are the same. If the commands are not the same, it is a violation. Figure 6-11 shows a sample waveform of this check in a simulation.

Marker 1 shows the point when a falling edge of the signal "framen" is detected. At this point a command "f7" is placed on the command bus. Marker 2 shows the point when the target device terminated the transaction by asserting the signal "stopn." The master makes another request and gets the grant for the bus. Marker 3 shows the point when the master asserts the signal "framen" again. At this point, a command of "f7" is placed on the bus once again and hence the check succeeds.

Master_chk9: Bus parity check errors for address phase (SERR), this can be checked for all the different types of transactions like memory read, memory write, I/O read, I/O write, etc.

```
property p_mchk9;
@(posedge clk)
  $fell (framen)   ##1
  (par ^ $past (^(ad[31:0]^cxben[3:0])) == 1) |->
    ##[1:5] $fell (serrn);
endproperty

a_mchk9: assert property(p_mchk9);
c_mchk9: cover property(p_mchk9);
```

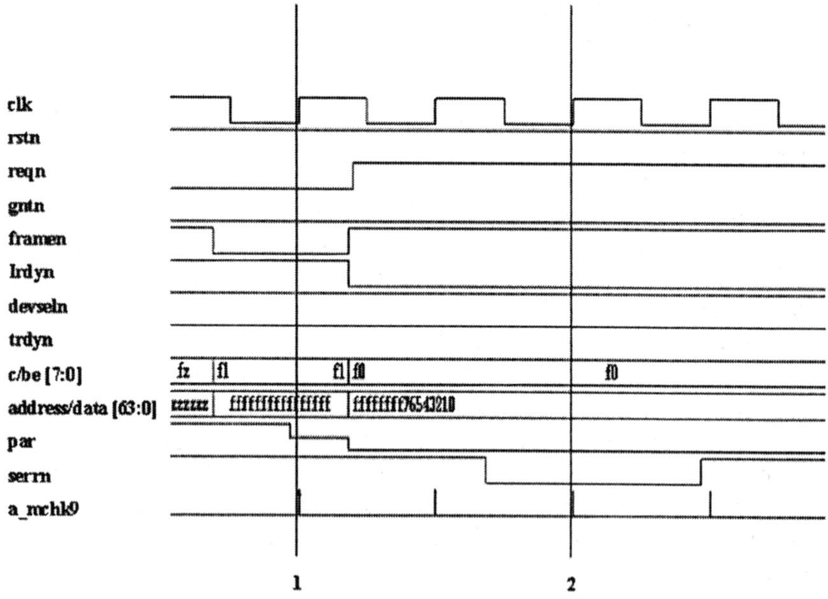

Figure 6-12. PCI Master check9

A parity check is performed during the address phase of every transaction. The parity should always be even for the signal "par" and the vectors "ad" and "c/be." This can be achieved by XOR'ing these three signals. Usually, the parity error is issued on the next clock cycle. If a parity error occurs during the address phase of a transaction, it indicates a system error and the signal "serrn" should be asserted. Figure 6-12 shows a sample waveform of this check in a simulation.

Marker 1 shows the point when the address phase is sampled. The value of the bus "c/be," value of the bus "address" and the signal "par" are sampled at this point and XOR'ed to find if even parity exists. Marker 2 shows that, even parity does not exist and hence the signal "serrn" is asserted. Note that the signal "serrn" is kept asserted for 2 clock cycles.

Master_chk10: Parity error in data phase (PERR).

A parity check is performed during the data phase of every transaction. The parity should always be even for the signal "par" and the vectors "ad" and "c/be." This can be achieved by XOR'ing these three signals. Usually, the parity error is issued on the next clock cycle. If a parity error occurs during the data phase of a transaction, the signal "perrn" should be asserted. Figure 6-13 shows a sample waveform of this check in a simulation.

```
property p_mchk10;
@(posedge clk)
(!irdyn && !trdyn) ##1
(par ^ $past (^(ad[31:0]^cxben[3:0])) == 1) |->
                          ##[1:5] !perrn;
endproperty

a_mchk10: assert property(p_mchk10);
c_mchk10: cover property(p_mchk10);
```

Marker 1 shows the point when the first data phase occurs out of the multiple data phases of this particular transaction. In the next clock cycle, the required parity value is set. This value is XOR'ed along with the value of the data and the command byte enable sampled from the previous clock cycle. If the parity bit is set incorrectly, the signal "perrn" should be asserted. For marker 1, the XOR'ed value of the data and command is 1 and the par bit is set to 1. Hence, there is no parity error. Marker 2 shows the second data phase. The XOR'ed value of the data (1234) and the command (0000) is 1 and the parity bit is set to 0. This does not provide even parity and hence the signal "perrn" is asserted in the next clock cycle, as shown by marker 3.

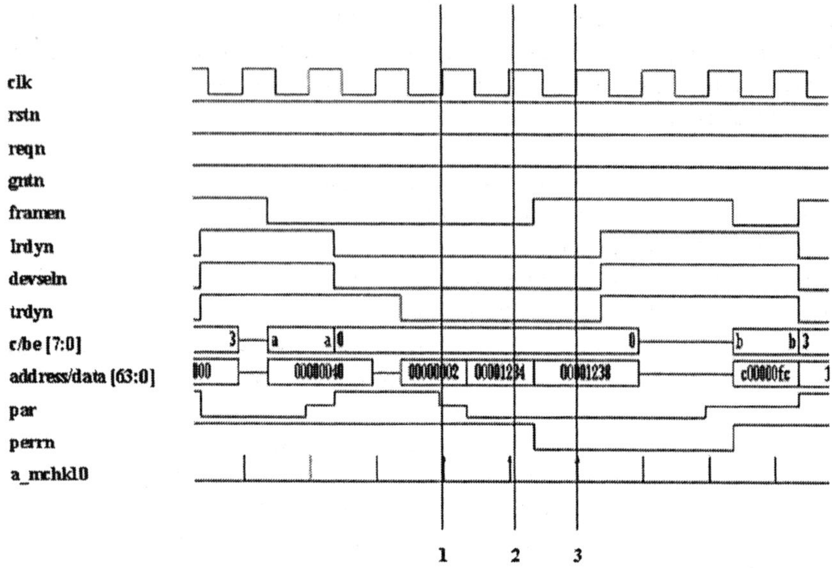

Figure 6-13. PCI Master check10

Master_chk11: PERR should not be asserted for special cycles.

The master can issue a command for special cycle. The command bus carries the value of "0001" during a special cycle. The parity error cannot be asserted during a special cycle irrespective of what data is driven into the data bus.

```
property p_mchk11;
@(posedge clk)
($fell (framen)&&(cxben[3:0] == (4'b0001))) |->
     (perrn [*1:$]
     ##0 ($rose (irdyn && trdyn))
     ##1 perrn[*2]);
endproperty

a_mchk11: assert property(p_mchk11);
c_mchk11: cover property(p_mchk11);
```

If the master asserts the signal "framen" during a special cycle, the signal "perrn" cannot be asserted until the bus becomes idle. Note that, it is necessary to make sure that the signal "perrn" is not asserted for 2 cycles even after the signals "trdyn" and "irdyn" are de-asserted. Since the parity error is issued in the next clock cycle of a data phase and the parity error is normally asserted for 2 clock cycles, this extension is necessary for the checker. Figure 6-14 shows a sample waveform of this check in a simulation. Marker 1 shows that the master has asserted the signal "framen" and issued the command "0001" on the c/be bus indicating that it is a special cycle. Note that the signal "perrn," which indicates a parity error, remains de-asserted irrespective of what the "par" bit value is. Hence, the checker succeeds. Marker 2 shows a similar special cycle.

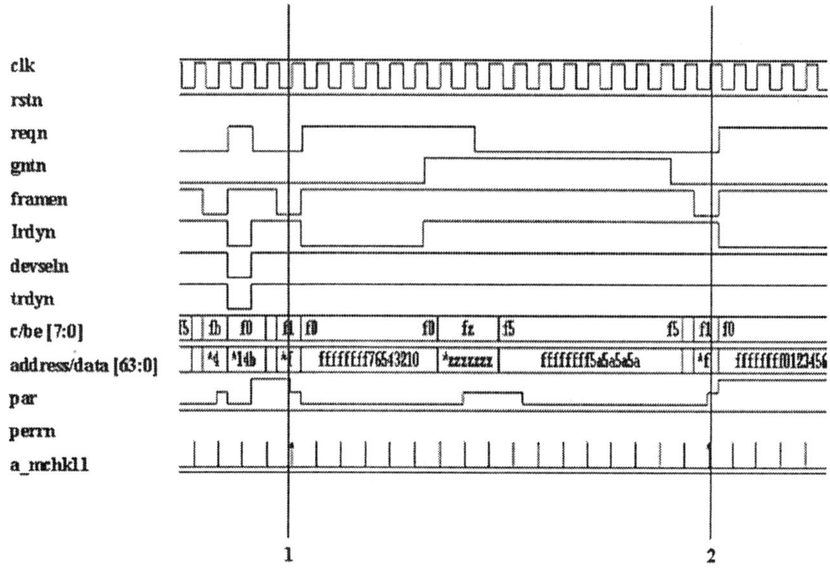

Figure 6-14. PCI Master check11

Master_Chk12: Dual Address Cycle.

It takes two cycles for the master to assert "irdyn" if it addresses a 64-bit target device. When the master asserts the signal "framen," it also issues the command for the dual address cycle. Along with the command, it also asserts the signal "req64n" to let the target know that the master wishes to perform a 64-bit transaction.

```
property p_mchk12;
@(posedge clk)
`s_DUAL_ADDR_CYCLE && req64n |=>
```

```
        not  $fell (irdyn);
endproperty

a_mchk12: assert property(p_mchk12);
c_mchk12: cover property(p_mchk12);
```

Master_chk13: Full 64-bit Transactions.

The master device asserts the signal "req64n" along with the "framen" signal to let the target device know that it wants to perform a 64-bit transaction. The target device responds by asserting the signals "devseln" and "ack64n" within 1 to 5 clock cycles.

```
property p_mchk13;
@(posedge clk)
$fell (gntn) ##[1:8]
$fell (framen) && $fell(req64n) |->
     ##[1:5] $fell (ack64n) && $fell(devseln);
endproperty

a_mchk13: assert property(p_mchk13);
c_mchk13: cover property(p_mchk13);
```

Figure 6-15 shows a sample waveform of this check in a simulation. Marker 1 shows the point when the signal "gntn" is asserted. In the next clock cycle, the master asserts the "framen" signal and the "req64n" signal, as shown by marker 2. This alerts the target device that the master wants to perform a 64-bit transaction. The target acknowledges the request of the master by asserting the signal "ack64n" along with the signal "devseln" in the next clock cycle. The master asserts the signal "irdyn" in the next clock cycle. Note that the master takes 2 clock cycles to assert the signal "irdyn" after asserting the "framen" signal in a 64-bit transaction.

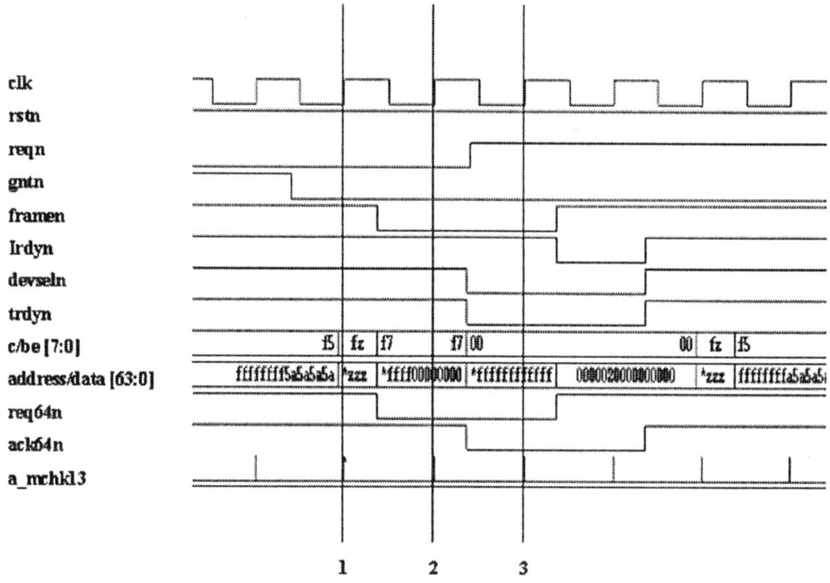

Figure 6-15. PCI Master check13

Master_Chk14: Check par64 signal validity.

Similar to the 32-bit transactions, a parity bit is used for 64-bit transactions. An even parity is maintained on the XOR'ed value of the most significant 32 bits of the data bus and the 4 most significant bits of the command byte enable, using the signal "par64."

```
property p_mchk14;
@(posedge clk)
(!ack64n && !irdyn && !trdyn && !devseln) &&
(^(ad[63:32]^cxben[7:4]) == 1) |=>
                              par64;
endproperty

a_mchk14: assert property(p_mchk14);
c_mchk14: cover property(p_mchk14);
```

Figure 6-16 shows a sample waveform of this check in a simulation. Marker 1 shows the point when a valid 64-bit data phase occurs. The XOR value of the 32 MSB of data (00000200) and the 4 MSB of c/be (0000) is 1. Hence, to maintain the even parity, the value of the signal "par64" should be a 1 in the next clock cycle. As seen in marker 2, the value of "par64" is detected as 1 and hence the check succeeds.

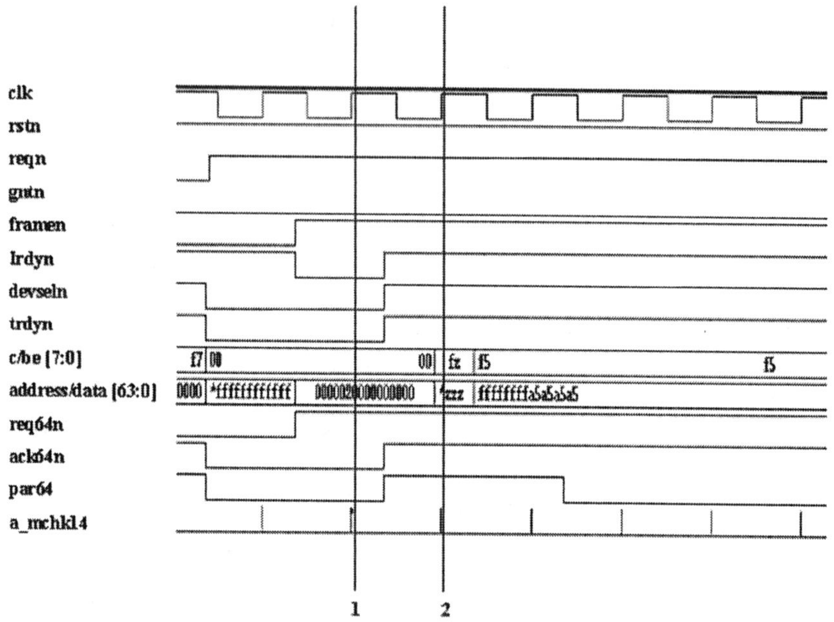

Figure 6-16. PCI Master check14

Master_chk15: Bus Parking

Bus parking happens when the "reqn" signal of a master is de-asserted but it still has the grant for the bus. The master goes to the idle state and then drives a stable value into the data bus and the command bus to indicate that it has parked the bus. When the master goes to the idle state, "reqn" should not be asserted. If "reqn" is asserted, it is considered as a back to back transaction.

```
sequence s_mchk15;
@(posedge clk)
first_match($fell (framen) ##[1:$]
      (framen && irdyn && !gntn && reqn));
endsequence

property p_mchk15;
```

```
@(posedge clk)
  s_mchk15 |->
  ##[1:8] (($stable(ad[31:0]))
 && ($stable (cxben[3:0])))
 ##1 (par ^ $past (^(ad[31:0]^cxben[3:0])) == 0);
endproperty

a_mchk15: assert property(p_mchk15);
c_mchk15: cover property(p_mchk15);
```

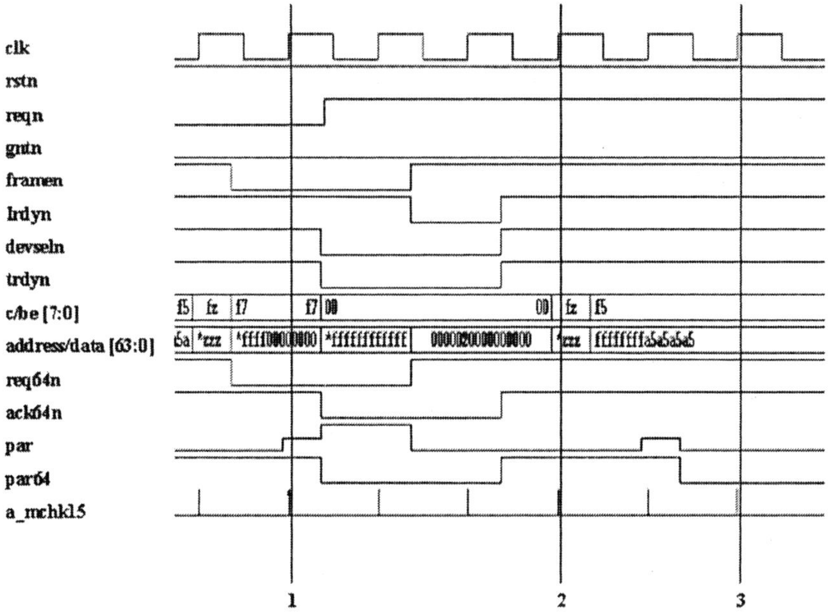

Figure 6-17. PCI Master check15

A simple sequence s_mchk15 is written to identify a valid completion of a master transaction. At the completion of the transaction, if the master still has the grant, then it is expected that it will drive a stable value into the data bus and the command bus, hence parking the PCI bus. The **$stable** function is used to detect if the bus values have stabilized. It is expected that one cycle later, the correct parity bit is set for the stable values. The XOR technique is used once again to detect the validity of the parity bit.

Figure 6-17 shows a sample waveform of this check in a simulation. Marker 1 shows a valid start of a transaction and marker 2 shows the completion of the transaction. Note that at this point, the signal "reqn" is de-asserted but the signal "gntn" is still asserted. Hence, the master is expected to drive a stable value into the data bus and command bus within 1 to 8 cycles. Marker 3 shows the point when the master has parked the bus.

Master_chk16: Fast back to back transactions.

A master device can perform fast back to back transactions wherein the signal "framen" is asserted in the immediate next clock cycle after the completion of a transaction. This can happen both at the completion of a single data phase or a multiple data phase sequence. Property p_mchk16 will capture only the single data phase transactions whereas property p_mchk17 will capture both single data phase and multiple data phase transactions.

```
property p_mchk16;
@(posedge clk)
($rose (framen) && $fell (irdyn))
##1 $fell (framen) |->
                $rose (irdyn);
endproperty

a_mchk16: assert property(p_mchk16);
c_mchk16: cover property(p_mchk16);


property p_mchk17;
@(posedge clk)
(!irdyn && framen)
##1 $fell (framen) |->
                 $rose (irdyn);
endproperty

a_mchk17: assert property(p_mchk17);
c_mchk17: cover property(p_mchk17);
```

Note that the main difference between the two properties is the sampling mechanism. If it is a single data phase back to back transaction, the signals "framen" and "irdyn" are sampled for their edges (falling edge of "framen" and rising edge of "irdyn"). Figure 6-18 shows a sample waveform of this check in a simulation.

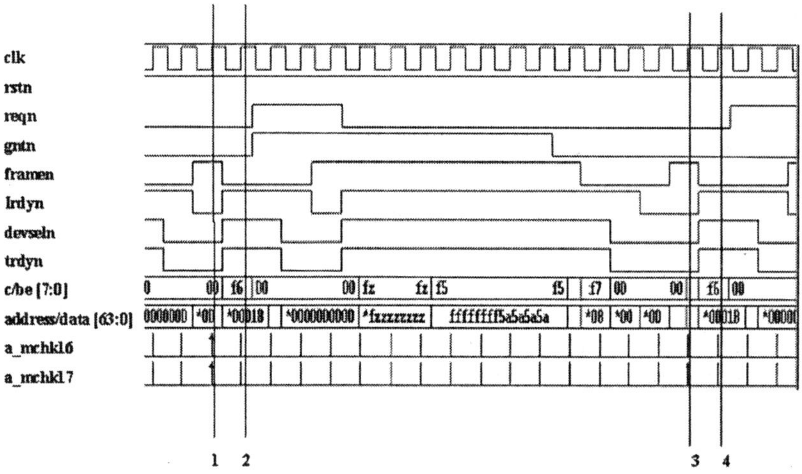

Figure 6-18. PCI Master check 16/17

Markers 1 and 2 indicate a single data phase back to back transaction. Markers 3 and 4 indicate a multiple data phase back to back transaction. Note that property p_mchk17 is sufficient to capture both scenarios.

Sample functional coverage point for PCI Master:

Master Abort - The master abort can happen in any of the following conditions - I/O read, I/O write, Configuration read, Configuration write, Memory read, Memory write. A cover statement can be written to make sure that the testbench executed all of these possible abort conditions at least once. After the master asserts the "irdyn" signal, if a target device does not respond within 5 clock cycles by asserting the "devseln" signal, the master will abort the transaction by de-asserting the "irdyn" signal. Note that the commands are specified in the properties by using the `define code.

```
property p_mcov1;
@(posedge clk)
`s_IO_READ ##1 (devseln)[*5] |=>
                        $rose (irdyn);
endproperty

property p_mcov2;
@(posedge clk)
`s_IO_WRITE ##1 (devseln)[*5] |=>
```

```
                          $rose (irdyn);
endproperty

property p_mcov3;
@(posedge clk)
`s_MEM_READ ##1 (devseln)[*5] |=>
                          $rose (irdyn);
endproperty

property p_mcov4;
@(posedge clk)
`s_MEM_WRITE ##1 (devseln)[*5] |=>
                          $rose (irdyn);
endproperty

property p_mcov5;
@(posedge clk)
`s_CONFIG_READ ##1 (devseln)[*5] |=>
                          $rose (irdyn);
endproperty


property p_mcov6;
@(posedge clk)
`s_CONFIG_WRITE  ##1 (devseln)[*5] |=>
                        $rose (irdyn);
endproperty

c_mcov1: cover property(p_mcov1);
c_mcov2: cover property(p_mcov2);
c_mcov3: cover property(p_mcov3);
c_mcov4: cover property(p_mcov4);
c_mcov5: cover property(p_mcov5);
c_mcov6: cover property(p_mcov6);
```

6.4 Scenario 2 – Target DUT Device

In this section, we assume that the design under test is a PCI target device. The rest of the system remains exactly the same. Figure 6-19 shows the sample system.

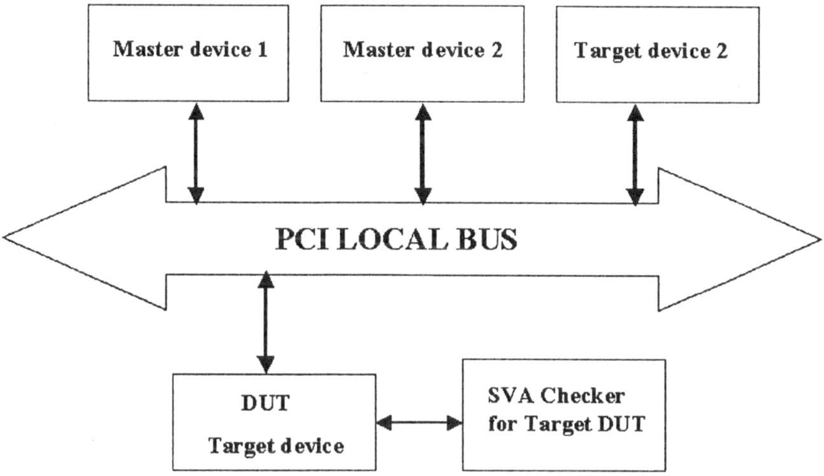

Figure 6-19. Sample Configuration for PCI Target device as DUT

6.4.1 PCI Target assertions

Target_chk1: Once target asserts the signal "stopn," it should keep "stopn" asserted until the signal "framen" is de-asserted, one clock cycle later "stopn" is de-asserted.

```
property p_tchk1;
@(posedge clk)
($fell (stopn) && !framen) |->
!stopn [*1:$]
##0 $rose (framen) ##1 $rose(stopn);
endproperty

a_tchk1: assert property(p_tchk1);
c_tchk1: cover property(p_tchk1);
```

Note that the property uses the "repeat until" construct to make sure that the signal "stopn" is kept asserted until the signal "framen" is de-asserted. Figure 6-20 shows a sample waveform of this check in a simulation.

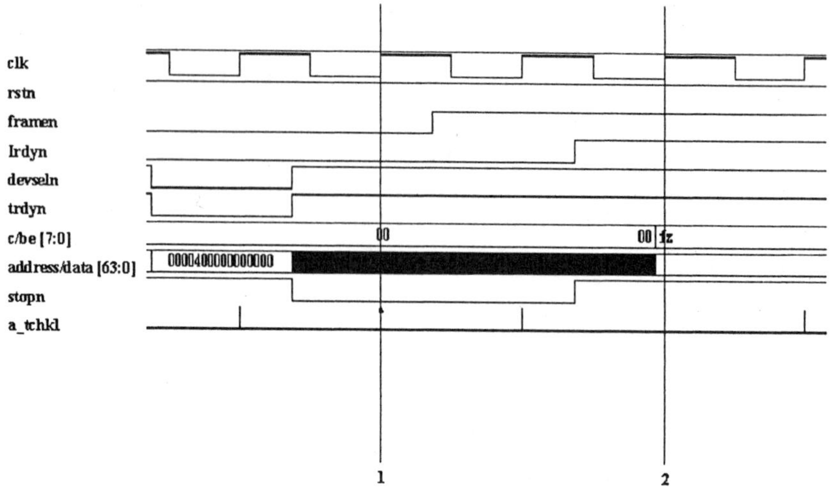

Figure 6-20. PCI Target check1

Marker 1 shows the point when the signal "stopn" is asserted. In the next clock cycle, the signal "framen" is de-asserted and one cycle later, the signal "stopn" is also de-asserted, as shown by marker 2.

Target_chk2: Once target has asserted the signal "trdyn," it cannot change "devseln" and "trdyn" until the current data phase completes.

When the target device asserts the signal "trdyn," it has acknowledged that it is ready to either accept data or send data. Hence, it cannot de-assert the signal "trdyn" without completing a data phase.

```
property p_tchk2;
@(posedge clk)
$fell (trdyn) |->
   (!trdyn && !devseln) [*0:16] ##0 !irdyn;
endproperty

a_tchk2: assert property(p_tchk2);
c_tchk2: cover property(p_tchk2);
```

The property p_tchk2 becomes active on the falling edge of the signal "trdyn." The consequent of the property makes sure that the signals "trdyn" and "devseln" stay asserted until the signal "irdyn" is asserted. The latency on the "trdyn" signal is 16 cycles.

Target_chk3: The target device cannot assert the signal "trdyn" until "devseln" is asserted.

```
property p_tchk3;
@(posedge clk)
     $fell (trdyn) |->!devseln;
endproperty

a_tchk3: assert property(p_tchk3);
c_tchk3: cover property(p_tchk3);
```

Target_chk5: Disconnect with data.

The target device indicates that it cannot continue a transaction by asserting the "stopn" signal and the "trdyn" signal at the same time. When this happens, the target is required to de-assert the "trdyn" signal in the next clock cycle but keep the signal "stopn" asserted. Hence, the last data phase completes without transferring any data since the signal "trdyn" is de-asserted. This is classified as "Disconnect – B" by the PCI local bus specification.

```
property p_tchk5b;
@(posedge clk)
($fell (stopn) && !framen && !trdyn && !irdyn)
 |=>
   (framen && trdyn)
   ##1 (stopn && devseln && irdyn);
endproperty

a_tchk5b: assert property(p_tchk5b);
c_tchk5b: cover property(p_tchk5b);
```

Figure 6-21 shows a sample waveform of this check in a simulation. Marker 1 shows the point when the signal "stopn" is asserted. In the next clock cycle, signal "trdyn" is de-asserted as expected. Marker 2 shows that, one clock cycle later, the signal "stopn," "devseln" and "irdyn" are all de-asserted hence completing the transaction.

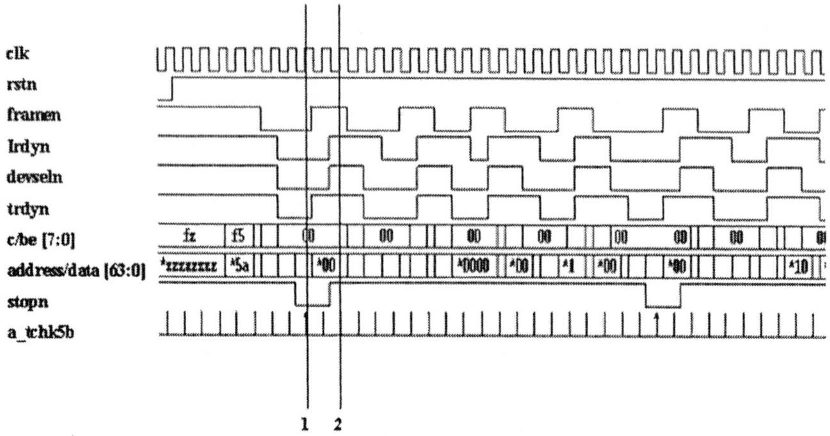

Figure 6-21. PCI Target check 5b

Target_chk6: Disconnect without data termination.

When the target device cannot complete any more data phases, it asserts the signal "stopn" and de-asserts the signal "trdyn." The target keeps the signal "stopn" asserted, until the final data phase is complete.

Note that a complex property like this should be split into smaller sequences as follows.

```
sequence s_tchk6a;
@(posedge clk)
(!irdyn && !trdyn && !devseln && !framen);
endsequence

sequence s_tchk6b;
@(posedge clk)
($fell (stopn) && $rose (trdyn) && !framen);
endsequence

sequence s_tchk6c;
@(posedge clk)
$rose (framen) ##[0:8] (!irdyn && !stopn);
endsequence

sequence s_tchk6;
```

```
@(posedge clk)
s_tchk6a.ended ##[1:8] s_tchk6b;
endsequence

property p_tchk6;
@(posedge clk)
  s_tchk6.ended |=> s_tchk6c;
endproperty

a_tchk6: assert property(p_tchk6);
c_tchk6: cover property(p_tchk6);
```

The sequence s_tchk6a identifies a valid data phase. The sequence s_tchkb identifies the point when the target device asserts the "stopn" signal. The sequence s_tchk6 is a concatenation of the two sequences s_tchk6a and s_tchk6b. The sequence s_tchk6 takes the check to the point, wherein a "stopn" has been issued. The sequence s_tchk6c looks for the point when both the signals "irdyn" and "stopn" are asserted. This is required because the master can get into a wait state before completing the last data phase, which therefore could have de-asserted the signal "irdyn."

Target_chk6_1: Master naturally terminating and target issuing an abort at the same time.

This is a case when the target is asserting the "stopn" signal to stop the transaction and at the same time the master device is also aborting the transaction naturally. This means that on the same clock cycle that the target asserts the "stopn" signal; the master de-asserts the "framen" signal.

```
sequence s_tchk6_1;
@(posedge clk)
(!irdyn && !trdyn && !devseln && !framen)
##[1:8] ($fell (stopn) && trdyn && framen);
endsequence

property p_tchk6_1;
@(posedge clk)
  s_tchk6_1.ended |=> (irdyn && stopn);
endproperty

a_tchk6_1: assert property(p_tchk6_1);
c_tchk6_1: cover property(p_tchk6_1);
```

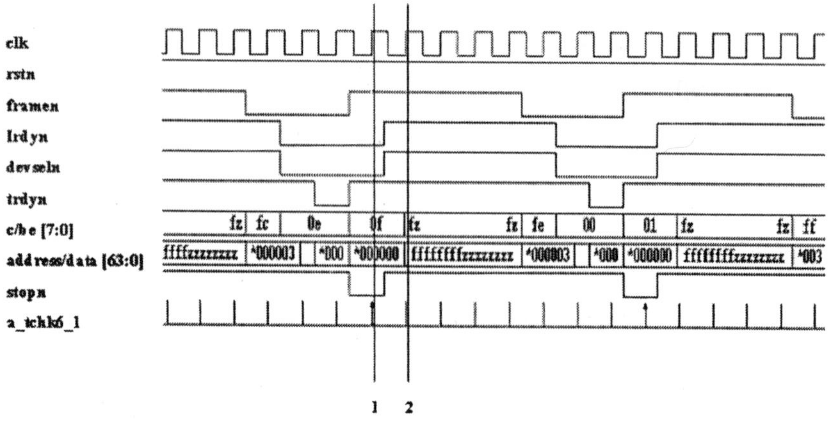

Figure 6-22. PCI Target check6_1

Note that a sequence is written to identify the point wherein both target and master are trying to stop the transaction simultaneously. The property checks that, if the antecedent matches, then the signal “irdyn” should be asserted in the next clock cycle along with the signal “stopn.” Figure 6-22 shows a sample waveform of this check in a simulation. Marker 1 shows the point when signal “framen” is de-asserted and the signal “stopn” is asserted. Marker 2 shows the point when the signals “irdyn” and “stopn” are de-asserted.

Target_chk7: Retry

If the target is not ready for a transaction, it has to ask the master to retry the transaction at a later point. This has to be done before the occurrence of the first data phase. The target device will assert the signal “stopn” before asserting the “trdyn” signal for the first time.

```
sequence s_tchk7a;
@(posedge clk)
$fell (framen) ##[1:8] $fell(irdyn);
endsequence

sequence s_tchk7b;
@(posedge clk)
$fell (framen) ##[1:5]
$fell(devseln) && $fell(stopn) && trdyn;
endsequence
```

```
sequence s_tchk7;
@(posedge clk)
  first_match(s_tchk7a and s_tchk7b);
endsequence

property p_tchk7;
  @(posedge clk) s_tchk7.ended |=> framen;
endproperty

a_tchk7: assert property(p_tchk7);
c_tchk7: cover property(p_tchk7);
```

The sequence s_tchk7 becomes active when the signal "framen" is asserted. Once the signal "framen" is asserted, it is expected that the target device identify itself by asserting the signal "devseln." It can happen anywhere between 1 to 5 clock cycles depending on the speed of the target device. If the target device wants to issue a retry, it will assert the signal "stopn" along with the signal "devseln." At this point, the signal "trdyn" should stay de-asserted. One cycle after the retry is issued, the master de-asserts the "framen" signal to acknowledge the retry issued by the target device.

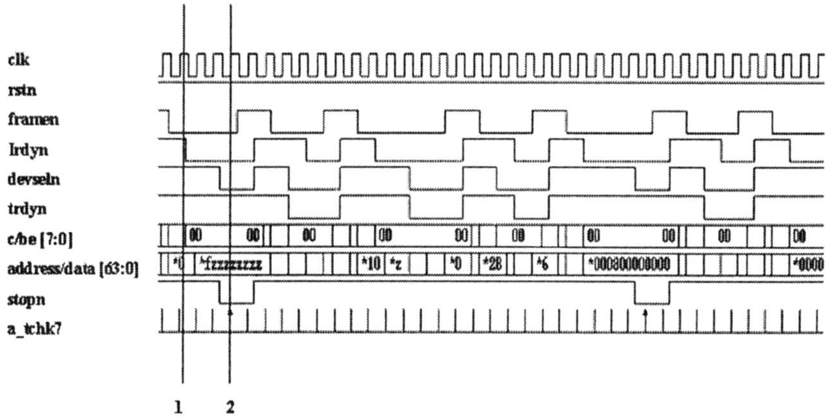

Figure 6-23. PCI Target check7

Figure 6-23 shows a sample waveform of this check in a simulation. Marker 1 shows the point when the master asserts the "framen" signal. Marker 2 shows the point when the target device asks the master to retry.

One cycle after marker 2, the master de-asserts the "framen" signal and ends the transaction.

Target_chk8: The signal "devseln" should not be asserted for a special cycle.

```
property p_tchk8;
@(posedge clk)
$fell (framen) && (cxben[3:0] == 4'b0001) |->
     devseln [*1:$] ##0 $rose (framen);
endproperty

a_tchk8: assert property(p_tchk8);
c_tchk8: cover property(p_tchk8);
```

During a special cycle, the signal "devseln" should not be asserted. The property becomes active when the "framen" signal is asserted and the master device places a special cycle command on the command bus. The consequent of the property makes sure that the signal "devseln" stays de-asserted until the master completes the transaction (by de-asserting the "framen" signal).

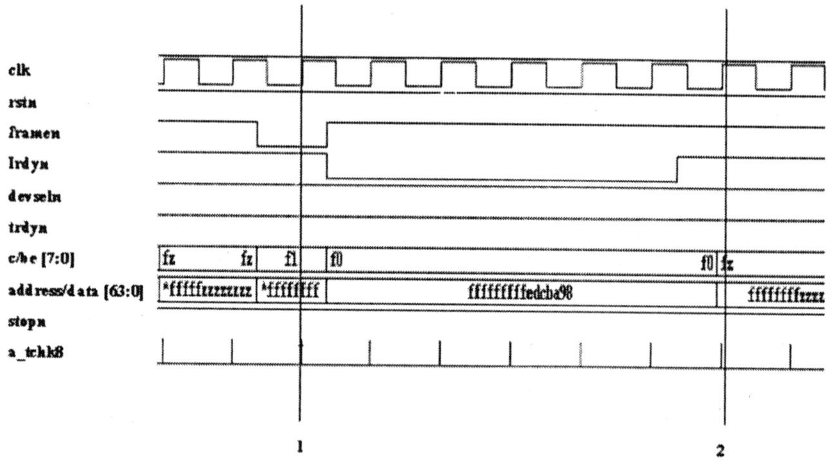

Figure 6-24. PCI Target check 8

Figure 6-24 shows a sample waveform of this check in a simulation. Marker 1 shows the point when a special cycle command is detected. Marker

2 shows the completion of the transaction. Note that from marker 1 to marker 2, the signal "devseln" stays de-asserted.

Target_chk9: Target latency for the completion of the first data phase is 16 cycles from the assertion of the signal "framen."

Once the master asserts the signal "framen," the target device identifies itself by asserting the signal "devseln" first. Depending on the nature of the target device, it can take anywhere from 1 to 5 cycles for the "devseln" signal to be asserted. For example, a fast target device takes only one clock cycle to respond, a medium target device takes 2 clock cycles to respond. After asserting the "devseln" signal, the target device will assert the "trdyn" signal if it is ready for a transaction. The total latency allowed by the PCI local bus specification, from the point the "framen" signal is asserted by the master to the point when an actual data phase happens (both "trdyn" and "irdyn" are asserted) is 16 clock cycles. Depending on the nature of the device the latency split can be summarized as shown in Table 6-2.

Two basic sequences are written to identify a valid data phase or a retry condition. A separate sequence is defined for each type of the target device. Note that the timing delay for the assertion of the "devseln" signal is the only difference between these sequences.

Table 6-2. Target latency table

Device type	Frame -> devsel	Devsel -> (irdy && trdy)
FAST	1	0:15
MEDIUM	2	0:14
SLOW	3	0:13
SUBTRACTIVE	4	0:12

```
sequence s_tchk9a;
@(posedge clk)
(!irdyn && !trdyn);
endsequence

sequence s_tchk9b;
@(posedge clk)
(!irdyn && !stopn);
endsequence

sequence s_tchk9_fast;
```

```
@(posedge clk)
$fell (framen) ##1 $fell(devseln);
endsequence
sequence s_tchk9_medium;
@(posedge clk)
$fell (framen) ##2 $fell(devseln);
endsequence

sequence s_tchk9_slow;
@(posedge clk)
$fell (framen) ##3 $fell(devseln);
endsequence

sequence s_tchk9_subtractive;
@(posedge clk)
$fell (framen) ##4 $fell(devseln);
endsequence

property p_tchk9_fast;
@(posedge clk)
s_tchk9_fast |-> ##[0:15]
   (!devseln) throughout
   (s_tchk9a.ended || s_tchk9b.ended);
endproperty

a_tchk9_fast: assert property(p_tchk9_fast);
c_tchk9_fast: cover property(p_tchk9_fast);

property p_tchk9_medium;
@(posedge clk)
s_tchk9_medium |-> ##[0:14]
   (!devseln) throughout
     (s_tchk9a.ended || s_tchk9b.ended);
endproperty

a_tchk9_medium: assert property(p_tchk9_medium);
c_tchk9_medium: cover property(p_tchk9_medium);

property p_tchk9_slow;
@(posedge clk)
s_tchk9_slow |-> ##[0:13]
   (!devseln) throughout
     (s_tchk9a.ended || s_tchk9b.ended);
```

```
endproperty

a_tchk9_slow: assert property(p_tchk9_slow);
c_tchk9_slow: cover property(p_tchk9_slow);

property p_tchk9_subtractive;
@(posedge clk)
s_tchk9_subtractive |-> ##[0:12]
   (!devseln) throughout
     (s_tchk9a.ended || s_tchk9b.ended);
endproperty

a_tchk9_subtractive:
     assert property(p_tchk9_subtractive);
c_tchk9_subtractive:
     cover property(p_tchk9_subtractive);
```

A separate property is written for each type of device. If the sequence mentioned in the antecedent of the property identifies a specific type of device, the consequent is allowed to take certain number of cycles to match, as specified in Table 6-2. For example, if it is a slow device, then the target device can take anywhere between 0 and 13 clock cycles to complete a valid data phase or issue a retry.

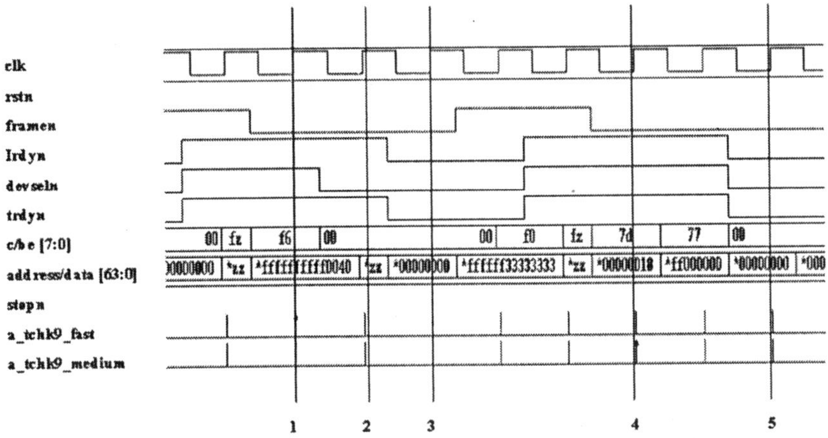

Figure 6-25. PCI Target check9

Figure 6-25 shows a sample waveform of this check in a simulation. Marker 1 shows the beginning of the property p_tchk9_fast. The master asserts the signal "framen" at this point. One cycle later, the signal "devseln" is asserted by the target device as shown by marker 2. One cycle after that, the signals "trdyn" and "irdyn" are asserted as shown by marker 3 and hence the check a_tchk9_fast succeeds.

Marker 4 shows the beginning of the property p_tchk9_medium. The master asserts the signal "framen" at this point. Two cycles later, the signal "devseln" is asserted by the target device as shown by marker 5. Note that the signals "trdyn" and "irdyn" are asserted on the same clock cycle and hence the check a_tchk9_medium succeeds.

Target_chk10: Latency for the subsequent data phase is 8 cycles from the previous data phase.

Both the master and target can issue wait states in between a transaction if they are not ready. If in a given clock edge, a data phase has just completed in a burst transaction, then the next data phase should occur within 8 clock cycles.

```
property p_tchk10;
@(posedge clk)
(!irdyn && !trdyn && !devseln && !framen) |->
    ##[1:8] (!irdyn && (!trdyn || !stopn));
endproperty

a_tchk10: assert property(p_tchk10);
c_tchk10: cover property(p_tchk10);
```

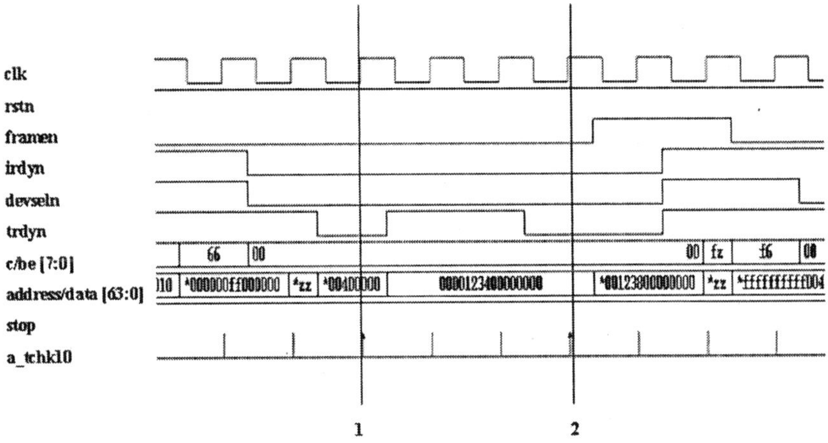

Figure 6-26. PCI Target check10

Figure 6-26 shows a sample waveform of this check in a simulation. Marker 1 shows a valid data phase. In the next clock cycle, the target device de-asserts the signal "trdyn" and hence issues a wait state. The wait state extends for one more cycle. One clock cycle after that, the signal "trdyn" is asserted again and hence a valid data phase occurs as shown by marker 2. In this case, the latency of the subsequent data phase is only 3 clock cycles and hence the check succeeds.

Target_chk11: The first data phase on a read command requires a turnaround cycle enforced by the signal "trdyn."

There are 4 possible read commands as shown in Table 6-1 and all read commands have a value of "10" in the 2 least significant bits. Whenever there is a read command, the master has to allow the target to drive the data into the bus and hence there is a turnaround cycle. The value of the data bus one clock cycle before the first data phase of a read cycle should be unknown.

```
sequence s_tchk11a;
@(posedge clk)
  ($fell (framen) && (cxben[1:0] == 2'b10));
endsequence

sequence s_tchk11b;
@(posedge clk)
```

```
first_match($fell (devseln) ##[1:16]
          $fell (trdyn));
endsequence

sequence s_tchk11;
@(posedge clk)
 s_tchk11a.ended ##[1:5] s_tchk11b;
endsequence

property p_tchk11;
@(posedge clk)
s_tchk11.ended |->
($isunknown (par)
&& $past ($isunknown(ad[31:0])));
endproperty

a_tchk11: assert property(p_tchk11);
c_tchk11: cover property(p_tchk11);
```

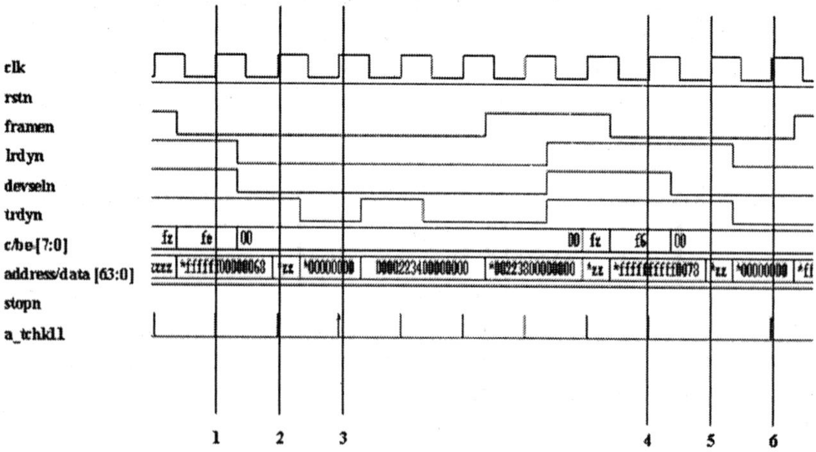

Figure 6-27. PCI Target check11

The sequence s_tchk11a detects a read command. The sequence s_tchk11b detects the first valid data phase after the read command was issued. The property p_tchk11 waits for the completion of the first data phase and then checks for the value of the par bit and the data bus in the previous cycle using the **$past** construct. If the values are not driven to "z" in the previous cycle, it is a violation.

Figure 6-27 shows a sample waveform of this check in a simulation. Marker 1 shows the point when a read command is detected (1110 – memory read line command). Marker 3 shows the point when the first valid data phase happens. One cycle before this, the data bus value should be a "z." Marker 2 shows that the value on the data bus is unknown and hence the check succeeds.

Marker 4 shows the point when a read command is detected (0110 – memory read command). Marker 6 shows the point when the first valid data phase happens. One cycle before this, the data bus value should be a "z." Marker 5 shows that the value on the data bus is unknown and hence the check succeeds.

Target_chk12: Configuration cycle (1).

During a valid configuration cycle, the 2 least significant bits of the address bus are set to either "00" or "01." When the configuration command is issued, the chip select signal "idsel" is asserted. The target device has to respond by asserting the signal "devseln" and eventually the configuration is completed when the signal "trdy" is asserted.

```
sequence s_tchk12a;
@(posedge clk)
(`s_CONFIG_READ || `s_CONFIG_WRITE) &&
((ad[1:0] == 2'b00) || (ad[1:0] == 2'b01)) &&
idsel;
endsequence

sequence s_tchk12b;
@(posedge clk)
!devseln && stopn;
endsequence

sequence s_tchk12;
@(posedge clk)
s_tchk12a ##[1:5] s_tchk12b;
endsequence

property p_tchk12;
@(posedge clk)
```

```
first_match(s_tchk12) |->
        ##[0:5] $fell (trdyn);
endproperty

a_tchk12: assert property(p_tchk12);
c_tchk12: cover property(p_tchk12);
```

Figure 6-28 shows a sample waveform of this check in a simulation. Marker 1 shows the point when a configuration command was issues by the system. Note that the signal "idsel" is asserted. Marker 2 shows the point when the target device asserts the signal "trdyn."

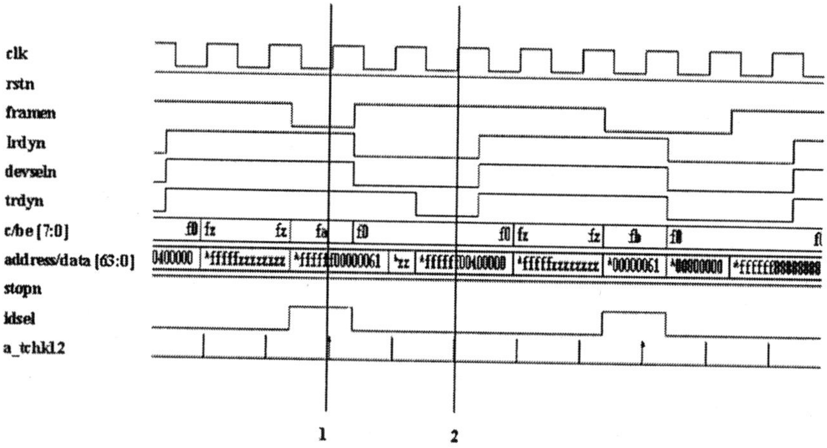

Figure 6-28. PCI Target check12

Target_chk13: Configuration cycle (2).

If the configuration command is issued and if the address bits are not set correctly ("10" or "11"), then the master should abort by de-asserting the "framen" signal.

```
sequence s_tchk13a;
@(posedge clk)
(`s_CONFIG_READ || `s_CONFIG_WRITE)
&& ((ad[1:0] == 2'b10) || (ad[1:0] == 2'b11))
&& idsel;
endsequence
```

```
sequence s_tchk13b;
@(posedge clk)
(devseln && stopn && trdyn) throughout
       (##[1:5] $rose (framen));
endsequence

property p_tchk13;
@(posedge clk)
   s_tchk13a |-> s_tchk13b;
endproperty

a_tchk13: assert property(p_tchk13);
c_tchk13: cover property(p_tchk13);
```

The sequence s_tchk13a detects an invalid configuration command. The sequence s_tchk13b makes sure that the signals "devseln," "trdyn" and "stopn" stay de-asserted until the signal "framen" is asserted. The signal "framen" should be de-asserted within 5 clock cycles.

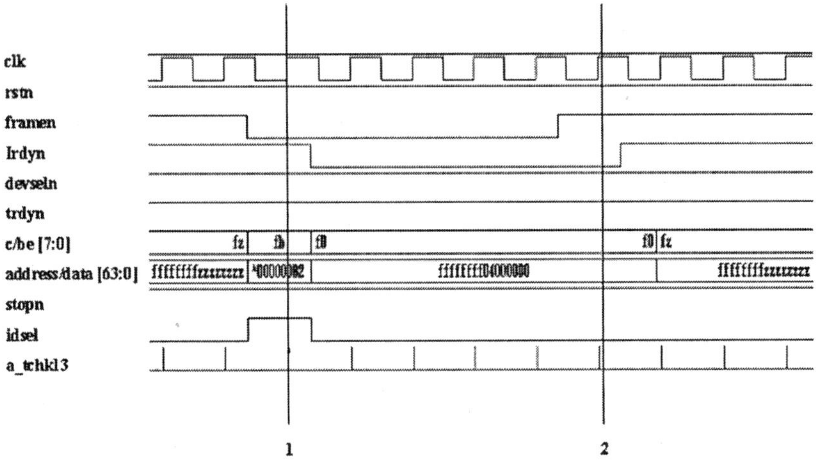

Figure 6-29. PCI Target check13

Figure 6-29 shows a sample waveform of this check in a simulation. Marker 1 shows the point when the invalid configuration command is detected. Marker 2 shows the point when the master aborts the transaction by de-asserting the "framen" signal. Note that the signals "trdyn," "devseln" and "stopn" stay de-asserted from marker1 to marker 2.

Sample functional coverage point for PCI Target:

Reserved commands – The signal "devseln" should not be asserted for any PCI reserved commands. The antecedent of the cover property looks for the reserved commands upon the assertion of the "framen" signal. The consequent makes sure that the signal "devseln" was kept de-asserted until the "framen" signal was de-asserted.

```
property p_tcov1;
@(posedge clk)
$fell (framen) && (cxben[3:0] == 4'b0100) |->
      devseln [*1:5] ##0 $rose (framen);
endproperty

c_tcov1: cover property(p_tcov1);

property p_tcov2;
@(posedge clk)
$fell (framen) && (cxben[3:0] == 4'b0101) |->
      devseln [*1:5] ##0 $rose (framen);
endproperty

c_tcov2: cover property(p_tcov2);

property p_tcov3;
@(posedge clk)
$fell (framen) && (cxben[3:0] == 4'b1000) |->
      devseln [*1:5] ##0 $rose (framen);
endproperty

c_tcov3: cover property(p_tcov3);

property p_tcov4;
@(posedge clk)
$fell (framen) && (cxben[3:0] == 4'b1001) |->
      devseln [*1:5] ##1 $rose (framen);
endproperty

c_tcov4: cover property(p_tcov4);
```

6.5 Scenario 3 – System level assertions

In this section, a few sample checks are shown for the PCI arbiter. The arbiter is usually part of the PCI bus. Figure 6-30 shows a sample system.

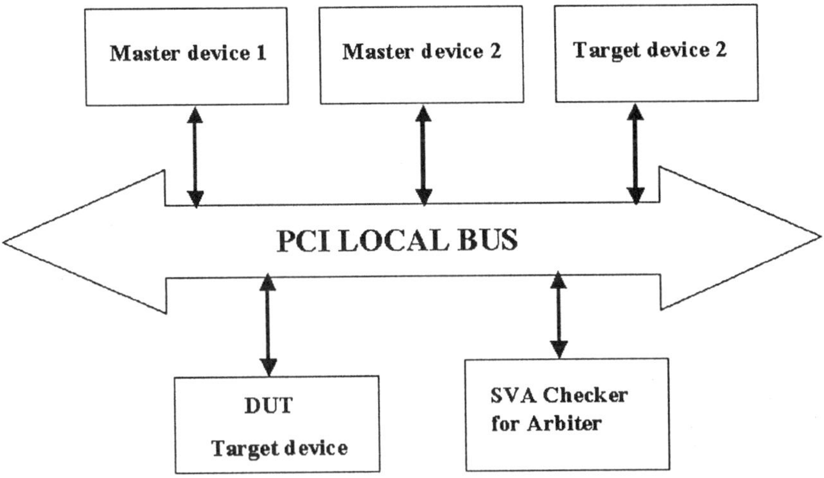

Figure 6-30. Sample PCI System for Arbiter checks

6.5.1 PCI Arbiter assertions

Arbiter_chk1: The signal "gntn" should be asserted when "framen" is asserted.

If the signal "gntn" is de-asserted and the signal "framen" is asserted in the same cycle, it is still valid.

```
property p_schk1;
@(posedge clk)
$fell (framen) |->
    !gntn[2] || $rose (gntn[2]);
endproperty

a_schk1: assert property(p_schk1);
c_schk1: cover property(p_schk1);
```

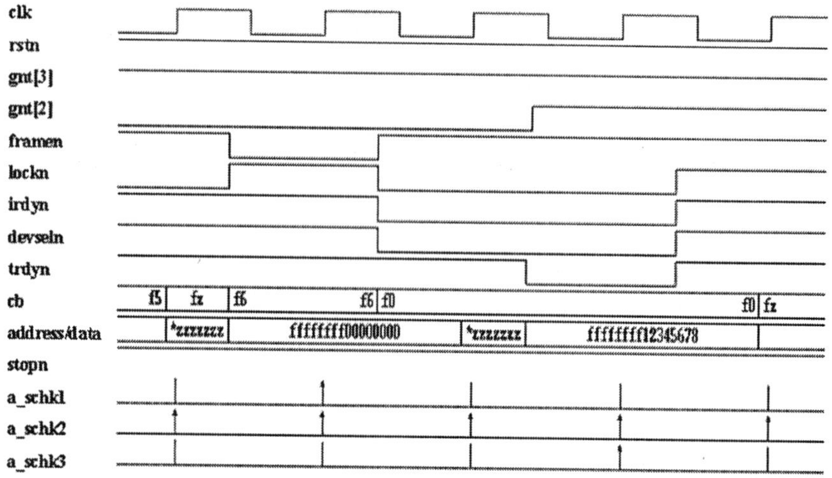

Figure 6-31. PCI Arbiter checks 1,2,3

Arbiter_chk2: Only one "gntn" signal can be asserted on a given clock cycle.

In the current sample system, there are two masters and hence, the arbiter uses two "gntn" signals.

```
property p_schk2;
@(posedge clk)
   $onehot0 ({!gntn[3], !gntn[2]});
endproperty

a_schk2: assert property(p_schk2);
c_schk2: cover property(p_schk2);
```

Since the "gntn" signals are active low signals, they are inverted and checked with a **zero one-hot** construct.

Arbiter_chk3: One "gntn" signal cannot be de-asserted and another asserted in the same cycle unless it is in idle cycle.

```
property p_schk3;
@(posedge clk)
$rose (gntn[2]) &&
(!framen || !irdyn) |->
```

```
        not $fell (gntn[3]);
    endproperty

    a_schk3: assert property(p_schk3);
    c_schk3: cover property(p_schk3);
```

Figure 6-31 shows a sample waveform of the checks a_schk1, a_schk2 and a_schk3 in a simulation.

Arbiter_chk4: The signal "lockn" should be asserted for the whole data phase.

```
sequence s_schk4a;
@(posedge clk)
first_match($fell (lockn) ##[0:5] !devseln);
endsequence

sequence s_schk4b;
@(posedge clk)
framen && !irdyn && (!trdyn || !stopn);
endsequence

property p_schk4;
@(posedge clk)
s_schk4a |-> !lockn [*1:$] ##0 s_schk4b;
endproperty

a_schk4: assert property(p_schk4);
c_schk4: cover property(p_schk4);
```

Arbiter_chk5: The signal "lockn" should be de-asserted during address phase.

```
property p_schk5;
@(posedge clk)
$fell (lockn) |->
     (($past (framen) == 0)
     && ($past(framen,2) == 1));
endproperty

a_schk5: assert property(p_schk5);
c_schk5: cover property(p_schk5);
```

The antecedent of the property looks for the assertion of the "lockn" signal. The address phase occurs when the "framen" signal is asserted. By checking for the falling edge of the "framen" signal, we can confirm that an address phase just occurred. The **$past** operator is used to get the value of the "framen" signal in the past two cycles.

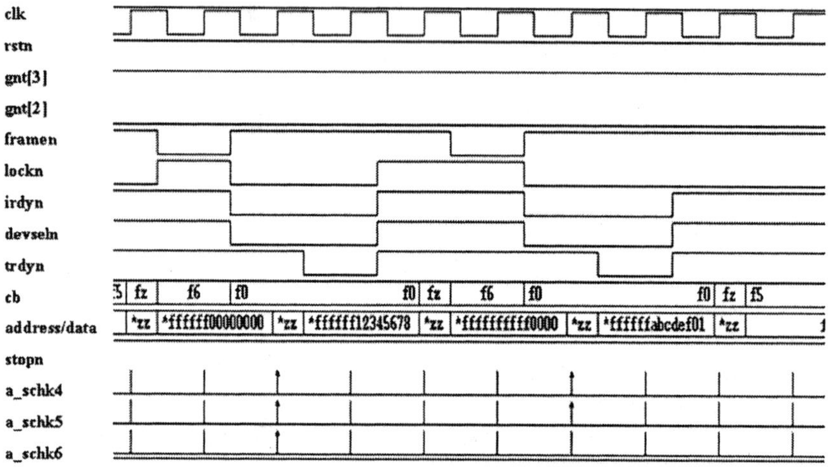

Figure 6-32. PCI Arbiter checks 4,5,6

Arbiter_chk6: The first transaction of a lock mechanism should be a read operation.

```
sequence s_schk6;
@(posedge clk)
first_match($fell (gntn[2]) ##[1:8]
        $fell (framen) ##1 $fell(lockn));
endsequence

property p_schk6;
@(posedge clk)
s_schk6.ended |->
    ($past(cxben[1:0]) == 2'b10);
endproperty

a_schk6: assert property(p_schk6);
c_schk6: cover property(p_schk6);
```

The sequence s_schk6 uses the "first_match" construct to identify the first assertion of the lock mechanism. The end of this is used as the antecedent of the property. The consequent part checks the 2 least significant bits of the command to make sure a read command was issued.

Figure 6-32 shows a sample waveform of the checks a_schk4, a_schk5 and a_schk6 in a simulation.

6.6 Summary on SVA for standard protocol

- Standard protocols are very complex and require a huge list of checkers to verify compliance.
- Timing rules are strict and these need to be achieved by the devices claiming to support these protocols.
- A common set of checkers can be developed for a particular interface and the same checkers can be re-used with other devices supporting similar interfaces.
- The complex nature of the protocol leads to multiple pre-conditions for most properties. Only SVA provides a variety of constructs and in-built mechanisms that can be used to define these complex pre-conditions.
- SVA also provides the capabilities to capture bus conditions using local variables. These local variables can be used effectively along with the pre-conditions to write complex temporal checks.
- SVA can be used effectively to create excellent functional coverage reports for a complex protocol.

Chapter 7

CHECKING THE CHECKER

Isolating assertion errors early

Assertion based verification provides excellent potential for finding design bugs early in the verification cycle. The SVA language is defined to address ABV with powerful built-in constructs. Assertion failures are indicated to the user by default as required by the SystemVerilog 3.1a standard. It is not required to display the success of an assertion by default. The user can use the action block of an assertion to display successes. Since the number of successes can be numerous (since most assertions are evaluated on every clock edge), displaying every success by default can create huge log files depending on the number of assertions that are active during simulation, slowing down the simulation.

A typical test configuration is shown in Figure 7-1. This is the same as Figure 0-2 shown in Chapter 0. Let's assume that a user executes this configuration and the simulation completes with a few assertion errors. The user should be absolutely confident that the error issued is a real design error. In other words, a user should be confident that his assertion code is correct and that the assertion failure is not a false condition. Debugging the entire design based on an assertion error is a tough task. If the error issued was due to bad assertion code, a user could waste a lot of time in the verification process. On the other hand, if there are no assertion failures during simulation, the verification engineer should be absolutely confident that the design works. If the assertion is not written accurately, it might not capture the intent of the design and hence, can miss a real error.

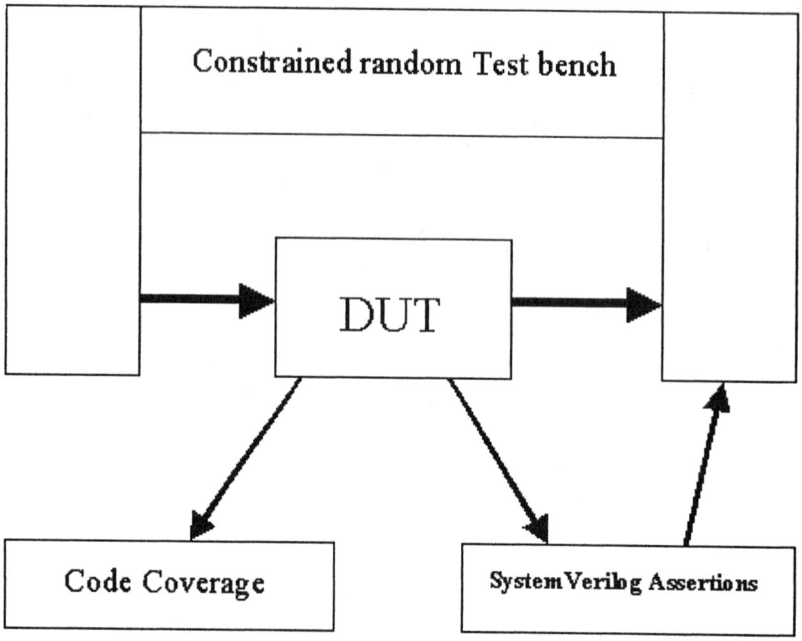

Figure 7-1. Typical simulation configuration

The declarative nature of the language makes SVA checks look very concise. If the checks are not coded well, the real intent of the property may not get represented accurately. It is critical to verify the functionality of the assertion code before binding it to the design. This involves investing some time upfront in the verification process, but it will prevent a user from navigating along a wrong debugging path. This chapter provides a few tips on how to check the checker. A sample configurable testbench for checking assertions that can be written between two signals is discussed in detail. The theory behind the configurable testbench will be used to verify a more complex protocol checker. Assertion validation is a vast topic and we just try exploring a few basic techniques in this chapter.

7.1 Assertion Verification

A simple testbench can be created to verify the functionality of an assertion. In most cases, the number of input conditions for an assertion is a finite number. This assumes that unbounded timing is not used in the definition of the property. An exhaustive testbench could be written to test all possible input conditions. If an unbounded timing is involved, it becomes impossible to create all possible input design conditions. Checkers involving unbounded timing can detect incorrect behavior, but do not fail at the end of simulation if an expected event does not occur. This is because the checker can not assume that a missing event would not happen if the simulation was run a bit longer. Unbounded timing checkers are thus considered incomplete. It is important to realize that even with a bounded time, the number of possible input design conditions can be numerous. This really depends on the complexity of the checker. An assertion is always based on two important concepts, as shown in Figure 7-2:

1. Logical relationship
2. Temporal relationship

If an assertion is written by logically combining (and, or, xor, etc.) an n-bit expression, then the possible number of input conditions is $(2^n - 1)$. Consider the logical expression shown below:

```
signal1 && signal2 && signal3
```

This expression will have 8 possible input conditions that need to be tested to guarantee the correct evaluation of the expression.

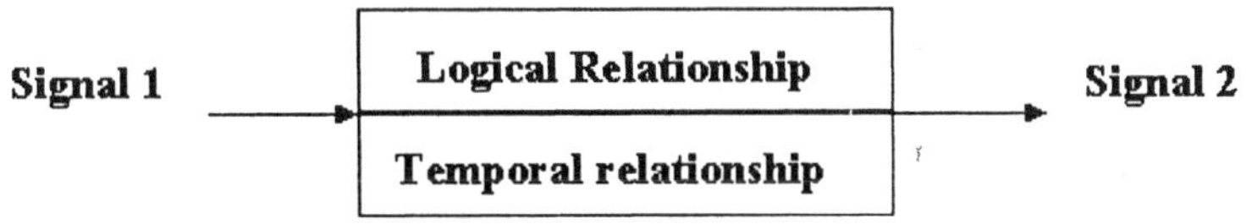

Figure 7-2. Assertion relationship

An assertion that involves a timing relationship between two signals can be tested thoroughly by using the bounds of the minimum and maximum timing limits. For example, consider the following case:

```
signal1 ##[min:max] signal2
```

This expression can fail on two conditions:

1. If the timing between the two signals is less than “min.” (This implies that signal 2 arrived before “min” time and did not stay true between “min” and “max” time)
2. If the timing between the two signals is more than “max.”

The assertion must succeed for all timings within the window (min and max). This means that varying the timing between signal1 and signal2 from (min-1) to (max+1) will cover all possible successes and at the least one error condition. If the timing between signal1 and signal2 is fixed, then the value of "min" and "max" are same. For a fixed timing relation, all possible successes and at least one error condition can be observed by varying the time from (min-1) to (min+1).

7.2 Assertion Test Bench (ATB) for SVA with two signals

In this section, we show how to create a configurable ATB. The objective is to create a testbench that can generate stimulus for SVA code that is written for two signals. The most basic requirement of an ATB is to get a correct response from the assertion for all possible successes and at least one error. An assertion involving two signals always has a leading signal (LS) and a trailing signal (TS). Consider the examples shown below:

```
signal1 && signal2
signal1 |-> signal2
signal1 ##[1:3] signal2
signal1 |-> ##2 signal2
```

In all these examples, irrespective of whether we are checking for a logical relationship or a timing relationship, we will address signal1 as the leading signal and signal2 as the trailing signal.

7.2.1 Logical relationship between two signals

There are several possible logical relationships between two signals involved in an SVA. Logical relationships are evaluated on a per clock basis. In other words, these are combinational checks.

Figure 7-3 shows a logical relationship tree for two signals. Based on the figure, there are 16 possible logical relationships between two signals. The logical relationship tree involves the following possibilities:

1. Logical relationship between two level sensitive signals.
2. Logical relationship between two edge sensitive signals.
3. Logical relationship between two level sensitive signals with overlapping implication.
4. Logical relationship between two edge sensitive signals with overlapping implication.

Any assertion that involves two level sensitive signals has the following possibilities for the leading signal (LS) and the trailing signal (TS):

a. HH – Both LS and TS are high
b. HL – LS is high and TS is low
c. LH – LS is low and TS is high
d. LL – Both LS and TS are low

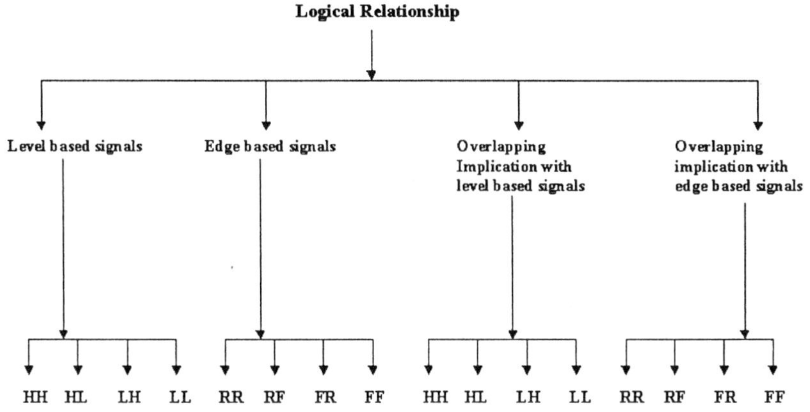

Figure 7-3. Logical relationship tree for SVA with two signals

Any assertion that involves two edge sensitive signals has the following possibilities for the leading signal (LS) and the trailing signal (TS):

a. RR – Both LS and TS have a rising edge
b. RF – LS has a rising edge and TS has a falling edge
c. FR – LS has a falling edge and TS has a rising edge
d. FF – Both LS and TS have falling edges

Note that the overlapping implication is listed as part of the logical relationship tree. If there is no timing involved between the leading signal and the trailing signal, the checker with an overlapping implication is a simple "if" statement. Hence, it can be grouped with the logical relationship tree.

It is possible to have a mix of a level sensitive signal and an edge sensitive signal in the same assertion. These combinations are not listed as part of the tree to simplify the discussion. In a checker, if the leading signal is level sensitive and the trailing signal is edge sensitive, it can be grouped with the level sensitive checks. Similarly, if the leading signal is edge sensitive and the trailing signal is level sensitive, the checker can be grouped with the edge sensitive checks. In the next section, we show how to generate stimulus that can verify all 16 possible relationships, as shown in Figure 7-3 thoroughly.

7.2.2 Stimulus generation for logical relationship – Level sensitive

A list of possible properties for logical relationship between two level sensitive signals is shown below. Note that a logical "and" operator is used in this example. This can be replaced with any other logical operator and the stimulus generation will remain exactly the same.

```
// On a given clock edge, both leading signal and
// trailing signal are high

property p_1_hh;
  @(posedge clk) a && b;
endproperty

// On a given clock edge, the leading signal is
// high and the trailing signal is low

property p_1_hl;
  @(posedge clk) a && !b;
endproperty

// On a given clock edge, the leading signal is
// low and the trailing signal is high

property p_1_lh;
  @(posedge clk) !a && b;
endproperty

// On a given clock edge, both leading signal and
// trailing signal are low

property p_1_ll;
  @(posedge clk) !a && !b;
endproperty

a_1_hh : assert property(p_1_hh);
a_1_hl : assert property(p_1_hl);
```

```
a_1_lh : assert property(p_1_lh);
a_1_ll : assert property(p_1_ll);
```

Overlapping implication is very similar to a simple logical operator with the only difference of the pre-condition. There is a hidden "if" statement that evaluates the trailing signal conditionally. If the leading signal is not true, then the property succeeds by default. A list of possible properties for logical relationship between two level sensitive signals with overlapping implication is shown below.

```
// on a given clock edge, if the leading signal
// is high, check that the trailing signal is
// also high

property p4_oli_hh;
 @(posedge clk) a |-> b;
endproperty

// on a given clock edge, if the leading signal
// is high, check that the trailing signal is
// low

property p4_oli_hl;
 @(posedge clk) a |-> !b;
endproperty

// on a given clock edge, if the leading signal
// is low, check that the trailing signal is
// high

property p4_oli_lh;
 @(posedge clk) !a |-> b;
endproperty

// on a given clock edge, if the leading signal
// is low, check that the trailing signal is
// low

property p4_oli_ll;
 @(posedge clk) !a |-> !b;
endproperty

a4_oli_hh: assert property(p4_oli_hh);
```

```
a4_oli_hl: assert property(p4_oli_hl);
a4_oli_lh: assert property(p4_oli_lh);
a4_oli_ll: assert property(p4_oli_ll);
```

To check a logical relation between two level sensitive signals, there are four possible input conditions (00, 01, 10, 11). By producing stimulus that covers all these four possible conditions, one can verify any logical operation between two level sensitive signals. The same stimulus will also satisfy the 4 conditions (HH, HL, LH, LL) shown in Figure 7-3.

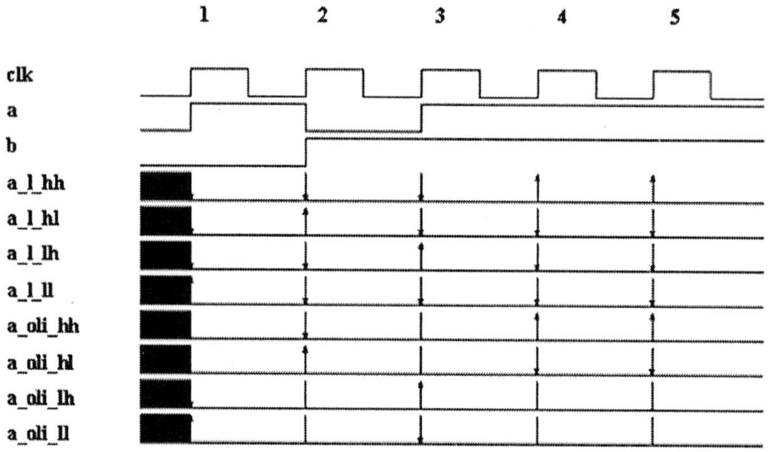

Figure 7-4. Waveform for logical relation between two level sensitive signals

The sample Verilog test code shown below is a simple 2-bit counter. The LSB of the counter drives the leading signal "a" and the MSB of the counter drives the trailing signal "b." The results produced by testing the above properties with the stimulus generated by the sample Verilog test code are shown in Figure 7-4. As shown in the figure, for the stimulus used, every assertion responded correctly for all real success and at least one real error.

```
// sample test code for logical relationship
// between level sensitive signals

logic [1:0] logical_op_reg;
logical_op_reg = 2'b00;

for(i=0; i<4; i++)
```

```
begin
 a <= logical_op_reg[0];
 b <= logical_op_reg[1];
 repeat(1) @(posedge clk);
 logical_op_reg++;
end
```

7.2.3 Stimulus generation for logical relationship – Edge sensitive

A list of possible properties for logical relationship between two edge sensitive signals is shown below.

```
// on a given clock edge the leading signal has a
// falling edge and the trailing signal has a
// falling edge

property p2_ff;
 @(posedge clk) $fell(a) && $fell(b);
endproperty

// on a given clock edge the leading signal has a
// falling edge and the trailing signal has a
// rising edge

property p2_fr;
 @(posedge clk) $fell(a) && $rose(b);
endproperty

// on a given clock edge the leading signal has a
// rising edge and the trailing signal has a
// falling edge

property p2_rf;
 @(posedge clk) $rose(a) && $fell(b);
endproperty

// on a given clock edge the leading signal has a
// rising edge and the trailing signal has a
// rising edge

property p2_rr;
 @(posedge clk) $rose(a) && $rose(b);
endproperty
```

```
a2_ff: assert property(p2_ff);
a2_fr: assert property(p2_fr);
a2_rf: assert property(p2_rf);
a2_rr: assert property(p2_rr);
```

A list of possible properties for logical relationship between two edge sensitive signals with overlapping implication is shown below.

```
// on a given clock edge, if the leading signal
// has a falling edge, then the trailing signal
// must have a falling edge

property p4_oei_ff;
 @(posedge clk) $fell(a) |-> $fell(b);
endproperty

// on a given clock edge, if the leading signal
// has a falling edge, then the trailing signal
// must have a rising edge

property p4_oei_fr;
 @(posedge clk) $fell(a) |-> $rose(b);
endproperty

// on a given clock edge, if the leading signal
// has a rising edge, then the trailing signal
// must have a falling edge

property p4_oei_rf;
 @(posedge clk) $rose(a) |-> $fell(b);
endproperty

// on a given clock edge, if the leading signal
// has a rising edge, then the trailing signal
// must have a rising edge

property p4_oei_rr;
 @(posedge clk) $rose(a) |-> $rose(b);
endproperty

a4_oei_ff: assert property(p4_oei_ff);
a4_oei_fr: assert property(p4_oei_fr);
a4_oei_rf: assert property(p4_oei_rf);
```

```
a4_oei_rr: assert property(p4_oei_rr);
```

Though we are checking only for logical relationship, the use of **$rose** or **$fell** constructs makes the stimulus generation slightly more complex. The stimulus generated should be capable of satisfying all the edge transitions. The sample Verilog test code shown below is a 2-bit counter and it accommodates all possible successes for both Fall-Fall and Fall-Rise conditions, as shown in the Figure 7-5.

```
// sample test code for logical relationship
// between edge sensitive signals

for(i=0; i<8; i++)
begin
  a <= logical_op_reg[0];
  b <= logical_op_reg[1];
  repeat(1) @(posedge clk);
  logical_op_reg++;
end
```

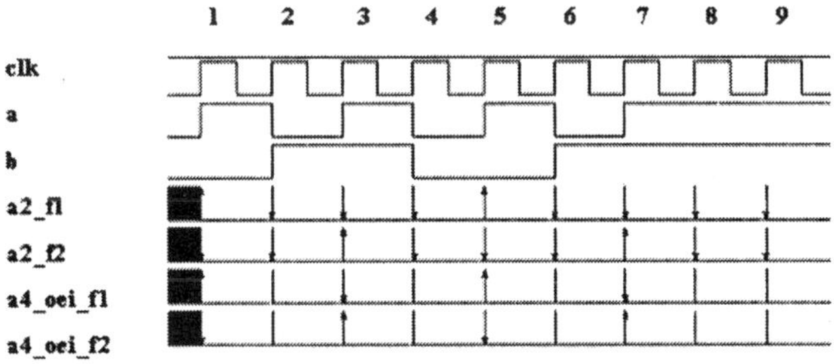

Figure 7-5. Logical condition on edge based signals - FF, FR

The sample Verilog test code shown below is also a 2-bit counter, but uses the negated values of the counter bits. It accommodates all possible successes for both Rise-Fall and Rise-Rise conditions, as shown in the Figure 7-6.

```
// sample test code for logical relationship
// between edge sensitive signals
for(i=0; i<8; i++)
begin
```

```
    a <= !logical_op_reg[0];
    b <= !logical_op_reg[1];
    repeat(1) @(posedge clk);
    logical_op_reg++;
  end
```

As seen in the last two sections, satisfying logical relationships is simple. The possible input conditions increase as the number of signals involved in the logical expression increase.

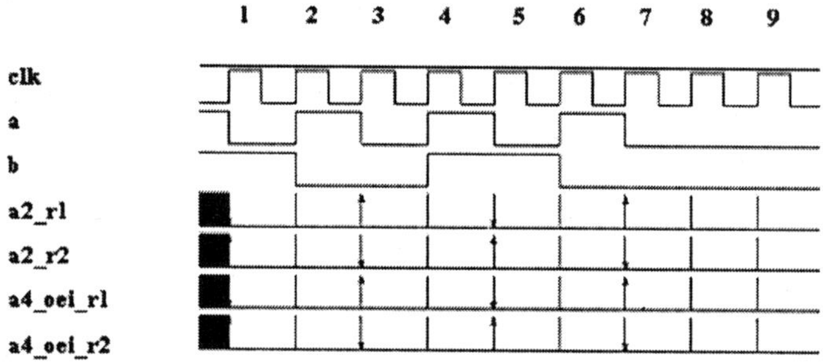

Figure 7-6. Logical condition on edge based signals - RR, RF

7.2.4 Timing relationship between two signals

The timing between the leading signal and the trailing signal can be a fixed delay or a variable delay. A timing relationship tree is very similar to the logic relationship tree except that it has twice the number of possibilities (fixed timing and variable timing) as shown in Figure 7-7. Also note that a timing relationship has a non-overlapping condition between the leading signal and the trailing signal.

The timing relationship tree involves the following possibilities:

1. Fixed timing relationship between two level sensitive signals.
2. Variable timing relationship between two level sensitive signals.
3. Fixed timing relationship between two edge sensitive signals.
4. Variable timing relationship between two edge sensitive signals.
5. Fixed timing relationship between two level sensitive signals with non-overlapping implication.

6. Variable timing relationship between two level sensitive signals with non-overlapping implication.
7. Fixed timing relationship between two edge sensitive signals with non-overlapping implication.
8. Variable timing relationship between two edge sensitive signals with non-overlapping implication.

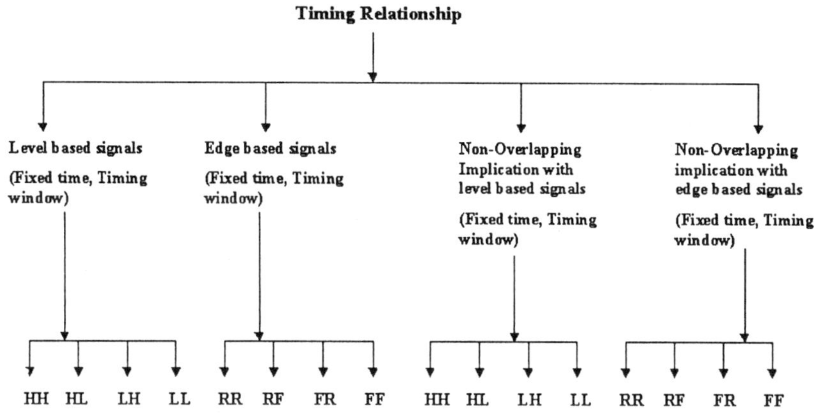

Figure 7-7. Timing relationship tree

A timing relationship between a level sensitive signal and an edge sensitive signal is possible. To simplify the timing relationship tree, these possibilities are not listed. As mentioned in Section 7.2.1, if the leading signal is level sensitive and the trailing signal is edge sensitive, it can be grouped with the level sensitive checks. Similarly, if the leading signal is edge sensitive and the trailing signal is level sensitive, the checker can be grouped with the edge sensitive checks.

7.2.5 Stimulus generation for timing relationship

A list of possible properties for fixed timing relationship between two level sensitive signals is shown below.

```
// On a given clock edge, the leading signal is
// high and after "min_time" clock cycles the
// trailing signal is high

property p3_hh;
  @(posedge clk) a ##min_time b;
```

```
endproperty

// On a given clock edge, the leading signal is
// high and after "min_time" clock cycles the
// trailing signal is low

property p3_hl;
 @(posedge clk) a ##min_time !b;
endproperty

// On a given clock edge, the leading signal is
// low and after "min_time" clock cycles the
// trailing signal is high

property p3_lh;
 @(posedge clk) !a ##min_time b;
endproperty

// On a given clock edge, the leading signal is
// low and after "min_time" clock cycles the
// trailing signal is low

property p3_ll;
 @(posedge clk) !a ##min_time !b;
endproperty

a3_f1: assert property(p3_hh);
a3_f2: assert property(p3_hl);
a3_f3: assert property(p3_lh);
a3_f4: assert property(p3_ll);
```

A list of possible properties for variable timing relationship between two level sensitive signals is shown below.

```
// On a given clock edge, the leading signal is
// high and between "min_time" and "max_time"
// clock cycles the trailing signal is high

property p3_w1_hh;
 @(posedge clk) a ## [min_time : max_time] b;
endproperty
// On a given clock edge, the leading signal is
// high and between "min_time" and "max_time"
```

```
// clock cycles the trailing signal is low

property p3_w2_hl;
 @(posedge clk) a ## [min_time : max_time] !b;
endproperty

// On a given clock edge, the leading signal is
// low and between "min_time" and "max_time"
// clock cycles the trailing signal is high

property p3_w3_lh;
 @(posedge clk) !a ## [min_time : max_time] b;
endproperty

// On a given clock edge, the leading signal is
// low and between "min_time" and "max_time"
// clock cycles the trailing signal is low

property p3_w4_ll;
 @(posedge clk) !a ## [min_time : max_time] !b;
endproperty

a3_w1: assert property(p3_w1_hh);
a3_w2: assert property(p3_w2_hl);
a3_w3: assert property(p3_w3_lh);
a3_w4: assert property(p3_w4_ll);
```

A list of possible properties for fixed timing relationship between two level sensitive signals with non-overlapping implication is shown below.

```
// On a given clock edge, if the leading signal
// is high, then after "min_time" clock cycles
// the trailing signal must be high

property p5_f_hh;
 @(posedge clk) a |-> ##min_time b;
endproperty

// On a given clock edge, if the leading signal
// is high, then after "min_time" clock cycles
// the trailing signal must be low
property p5_f_hl;
 @(posedge clk) a |-> ##min_time !b;
```

```
endproperty

// On a given clock edge, if the leading signal
// is low, then after "min_time" clock cycles
// the trailing signal must be high

property p5_f_lh;
 @(posedge clk) !a |-> ##min_time b;
endproperty

// On a given clock edge, if the leading signal
// is low, then after "min_time" clock cycles
// the trailing signal must be low

property p5_f_ll;
 @(posedge clk) !a |-> ##min_time !b;
endproperty

a5_f_hh: assert property(p5_f_hh);
a5_f_hl: assert property(p5_f_hl);
a5_f_lh: assert property(p5_f_lh);
a5_f_ll: assert property(p5_f_ll);
```

A list of possible properties for variable timing relationship between two level sensitive signals with non-overlapping implication is shown below.

```
// On a given clock edge, if the leading signal
// is high, then between "min_time" and
// "max_time" clock cycles the trailing signal
// must be high

property p5_w_hh;
 @(posedge clk)
 a |-> ##[min_time : max_time] b;
endproperty

// On a given clock edge, if the leading signal
// is high, then between "min_time" and
// "max_time" clock cycles the trailing signal
// must be low

property p5_w_hl;
 @(posedge clk)
```

```
  a |-> ##[min_time : max_time] !b;
endproperty

// On a given clock edge, if the leading signal
// is low, then between "min_time" and
// "max_time" clock cycles the trailing signal
// must be high

property p5_w_lh;
  @(posedge clk)
  !a |-> ##[min_time : max_time] b;
endproperty

// On a given clock edge, if the leading signal
// is low, then between "min_time" and
// "max_time" clock cycles the trailing signal
// must be low

property p5_w_ll;
  @(posedge clk)
  !a |-> ##[min_time : max_time] !b;
endproperty

a5_w_hh: assert property(p5_w_hh);
a5_w_hl: assert property(p5_w_hl);
a5_w_lh: assert property(p5_w_lh);
a5_w_ll: assert property(p5_w_ll);
```

A list of possible properties for fixed timing relationship between two edge sensitive signals is shown below.

```
// on a given clock edge, the leading signal has
// a falling edge and after "min_time" cycle the
// trailing signal has a falling edge

property p4_f_ff;
  @(posedge clk) $fell(a) ##min_time $fell(b);
endproperty

// on a given clock edge, the leading signal has
// a rising edge and after "min_time" cycle the
// trailing signal has a rising edge
```

```
property p4_f_rr;
 @(posedge clk) $rose(a) ##min_time $rose(b);
endproperty

// on a given clock edge, the leading signal has
// a falling edge and after "min_time" cycle the
// trailing signal has a rising edge

property p4_f_fr;
 @(posedge clk) $fell(a) ##min_time $rose(b);
endproperty

// on a given clock edge, the leading signal has
// a rising edge and after "min_time" cycles the
// trailing signal has a falling edge

property p4_f_rf;
 @(posedge clk) $rose(a) ##min_time $fell(b);
endproperty

a4_f_rr: assert property(p4_f_rr);
a4_f_ff: assert property(p4_f_ff);
a4_f_rf: assert property(p4_f_rf);
a4_f_fr: assert property(p4_f_fr);
```

A list of possible properties for variable timing relationship between two edge sensitive signals is shown below.

```
// on a given clock edge, the leading signal has
// a falling edge and within "min_time" to
// "max_time" cycles the trailing signal has a
// falling edge

property p4_w_ff;
 @(posedge clk)
 $fell(a) ##[min_time : max_time] $fell(b);
endproperty

// on a given clock edge, the leading signal has
// a rising edge and within "min_time" to
// "max_time" cycles the trailing signal has a
// rising edge
```

```
property p4_w_rr;
 @(posedge clk)
 $rose(a) ##[min_time : max_time] $rose(b);
endproperty

// on a given clock edge, the leading signal has
// a falling edge and within "min_time" to
// "max_time" cycles the trailing signal has a
// rising edge

property p4_w_fr;
 @(posedge clk)
 $fell(a) ##[min_time : max_time] $rose(b);
endproperty

// on a given clock edge, the leading signal has
// a rising edge and within "min_time" to
// "max_time" cycles the trailing signal has a
// falling edge

property p4_w_rf;
 @(posedge clk)
 $rose(a) ##[min_time : max_time] $fell(b);
endproperty

a4_w_rr: assert property(p4_w_rr);
a4_w_ff: assert property(p4_w_ff);
a4_w_rf: assert property(p4_w_rf);
a4_w_fr: assert property(p4_w_fr);
```

A list of possible properties for fixed timing relationship between two edge sensitive signals with non-overlapping implication is shown below.

```
// on a given clock edge, if the leading signal
// has a falling edge, then after "min_time"
// cycles the trailing signal must have a
// falling edge

property p6_f_ff;
 @(posedge clk)
 $fell(a) |-> ##min_time $fell(b);
endproperty
```

```
// on a given clock edge, if the leading signal
// has a rising edge, then after "min_time"
// cycles the trailing signal must have a
// rising edge

property p6_f_rr;
  @(posedge clk)
  $rose(a) |-> ##min_time $rose(b);
endproperty

// on a given clock edge, if the leading signal
// has a falling edge, then after "min_time"
// cycles the trailing signal must have a
// rising edge

property p6_f_fr;
  @(posedge clk)
  $fell(a) |-> ##min_time $rose(b);
endproperty

// on a given clock edge, if the leading signal
// has a rising edge, then after "min_time"
// cycles the trailing signal must have a
// falling edge

property p6_f_rf;
  @(posedge clk)
  $rose(a) |-> ##min_time $fell(b);
endproperty
a6_f_rr: assert property(p6_f_rr);
a6_f_ff: assert property(p6_f_ff);
a6_f_rf: assert property(p6_f_rf);
a6_f_fr: assert property(p6_f_fr);
```

A list of possible properties for variable timing relationship between two edge sensitive signals with non-overlapping implication is shown below.

```
// on a given clock edge, if the leading signal
// has a falling edge, then within "min_time" to
// "max_time" cycles the trailing signal must
// have a falling edge
property p6_w_ff;
  @(posedge clk)
```

```
  $fell(a) |-> ##[min_time : max_time] $fell(b);
endproperty

// on a given clock edge, if the leading signal
// has a rising edge, then within "min_time" to
// "max_time" cycles the trailing signal must
// have a rising edge

property p6_w_rr;
  @(posedge clk)
  $rose(a) |-> ##[min_time : max_time] $rose(b);
endproperty

// on a given clock edge, if the leading signal
// has a falling edge, then within "min_time" to
// "max_time" cycles the trailing signal must
// have a rising edge

property p6_w_fr;
  @(posedge clk)
  $fell(a) |-> ##[min_time : max_time] $rose(b);
endproperty

// on a given clock edge, if the leading signal
// has a rising edge, then within "min_time" to
// "max_time" cycles the trailing signal must
// have a falling edge

property p6_w_rf;
  @(posedge clk)
  $rose(a) |-> ##[min_time : max_time] $fell(b);
endproperty

a6_w_rr: assert property(p6_w_rr);
a6_w_ff: assert property(p6_w_ff);
a6_w_rf: assert property(p6_w_rf);
a6_w_fr: assert property(p6_w_fr);
```

The following sample Verilog test code can generate stimulus that will satisfy all timing relationships defined between two signals except eventuality. Note that the signals "a" and "b" are initialized based on what the "timing level" is set to. For example, if the "timing level" is set to a "10," then the test code will generate stimulus for an "active high" level on the leading signal and an "active low" level on the trailing signal, if it is a level sensitive checker. For an edge sensitve checker, if the "timing level" is set to a "10," then the test code will generate stimulus for a "rising edge" on the leading signal and a "falling edge" on the trailing signal. A simple "for" loop is used to generate timing windows starting from (min_time-1) to (max_time+3). This will make sure that all possible successes are created and at least one error is created. The same results can be achieved with an upper window of (max_time+1). By using (max_time+3), we produce more error conditions. If the timing is fixed, then the values of "min_time" and "max_time" are the same.

Figure 7-8 shows a sample waveform of a fixed timing relationship property between two level sensitive signals (leading signal is active low and trailing signal is active high). Figure 7-9 shows a sample waveform of a variable timing relationship property between two edge sensitive signals (leading signal and the trailing signal look for a rising edge).

```
// sample Verilog test code for timing
// relationship between two signals

if(timing_level == 2'b11) begin
a = 1'b0; b=1'b0; end
if(timing_level== 2'b00) begin
a = 1'b1; b=1'b1; end
if(timing_level == 2'b01) begin
a = 1'b1; b=1'b0; end
if(timing_level == 2'b10) begin
a = 1'b0; b=1'b1; end

for(i=(min_time-1); i<(max_time+3); i++)
begin
  repeat(1) @(posedge clk);
  a <= ~a;
  if(i == 0)
  begin
    b <= ~b;
    repeat(1) @(posedge clk);
    a <= ~a; b <= ~b;
  end
  else
  begin
    repeat(1) @(posedge clk);
    a<= ~a;
```

```
      repeat((i-1)) @(posedge clk);
      b<= ~b;
      repeat(1) @(posedge clk);
      b<= ~b;
    end
  end
```

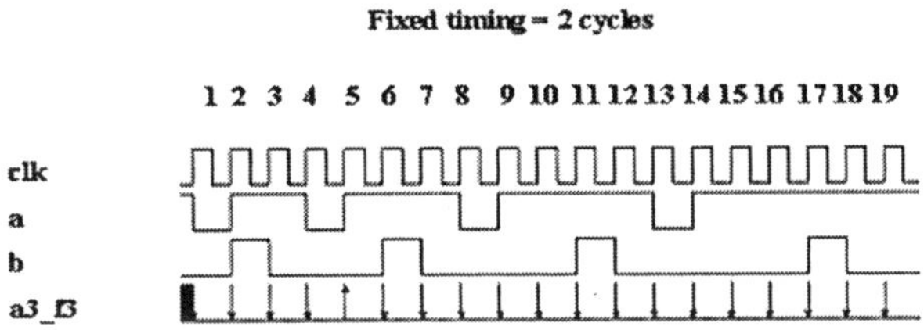

Figure 7-8. Timing (fixed) between two level sensitive signals

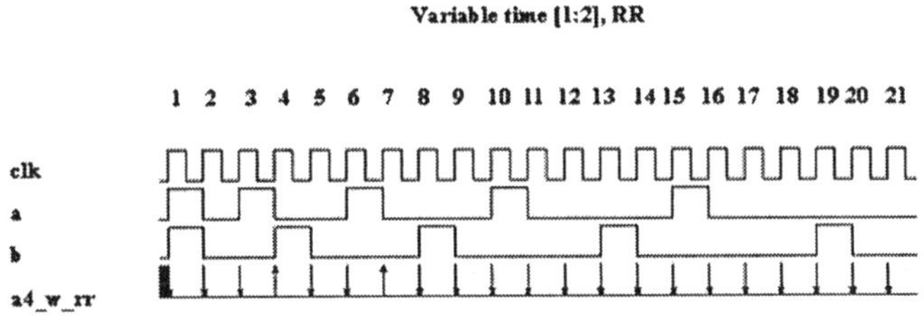

Figure 7-9. Timing (variable) between two edge sensitive signals

Timing relationships can be tested by looping from (min-1) to (max+1) values. This will guarantee that all possible successes are tested and at least one failure is tested. Repetition of signals is an extension of timing relationships. Repetitions also involve timing relationships, but it requires that the leading signal or the trailing signal repeat its value for a defined number of cycles. Repetition relationships are discussed in detail in the next section.

7.2.6 Repetition relationship between two signals

There are two main categories of repetition between two signals:

1. **Repeat after** – After an expected edge on the leading signal, with or without a time delay, the trailing signal is expected to repeat "n" times.

2. **Repeat until** – After an expected edge on the leading signal, the leading signal repeats until the expected value arrives on the trailing signal.

The repeat operators often involve the combination of an edge sensitive signal and a level sensitive signal. The leading signal is often an edge sensitive signal and the trailing signal is often a level sensitive signal. The "repeat until" condition can have an edge sensitive signal as a trailing signal. The naming convention for the repetition properties is as follows:

RH – LS has a rising edge and TS is high
RL – LS has a rising edge and TS is low
FH – LS has a falling edge and TS is high
FL – LS has a falling edge and TS is low

A list of possible properties for "repeat after" relationship between two signals is shown below.

```
// on a given clock edge the leading signal has a
// rising edge and after "start_wait" cycles, the
// trailing signal is high "repetition" times

property p7_c_rpt_rh;
  @(posedge clk)
  $rose(a) ##start_wait b[*repetition];
endproperty

// on a given clock edge the leading signal has a
// rising edge and after "start_wait" cycles, the
// trailing signal is low "repetition" times

property p7_c_rpt_rl;
  @(posedge clk)
  $rose(a) ##start_wait !b[*repetition];
endproperty

// on a given clock edge the leading signal has a
// falling edge and after "start_wait" cycles,
// the trailing signal is high "repetition" times

property p7_c_rpt_fh;
  @(posedge clk)
  $fell(a) ##start_wait b[*repetition];
endproperty
```

```
// on a given clock edge the leading signal has a
// falling edge and after "start_wait" cycles,
// the trailing signal is low "repetition" times

property p7_c_rpt_fl;
  @(posedge clk)
  $fell(a) ##start_wait !b[*repetition];
endproperty
```

A list of possible properties for "repeat until" relationship between two signals is shown below.

```
// on a give clock edge, the leading signal has a
// rising edge and stays high until the trailing
// signal is low

property p7_cu_rpt_rl;
  @(posedge clk) $rose(b) ##0 b[*1:$] ##1 !a;
endproperty

// on a given clock edge, the leading signal has
// a falling edge and stays low until the
// trailing signal is low

property p7_cu_rpt_fl;
  @(posedge clk) $fell(b) ##0 !b[*1:$] ##1 !a;
endproperty

// on a given clock edge, the leading signal has
// a rising edge and stays high until the
// trailing signal is high

property p7_cu_rpt_rh;
  @(posedge clk) $rose(b) ##0 b[*1:$] ##1 a;
endproperty

// on a given clock edge, the leading signal has
// a falling edge and stays high until the
// trailing signal is high

property p7_cu_rpt_fh;
  @(posedge clk) $fell(b) ##0 !b[*1:$] ##1 a;
endproperty
```

A sample code that can generate stimulus to verify both the "repeat after" and "repeat until" conditions for all possible properties is shown below. The stimulus uses the same concept as that of the timing. It loops between (repetition–1) and (repetition+3) to cover all possible successes and at the least one error condition. The variable "start_wait" defines the number of cycles to wait before looking for the repetition on the trailing signal (relevant to the "repeat_after" condition). The variable "stop_wait" defines the number of cycle to wait before re-setting the leading signal (relevant to the "repeat_until" condition).

```
// sample Verilog test code for repetition

logic [1:0] stop_wait;

if(rpt_edge == 2'b11) begin
a = 1'b0; b=1'b0; end

if(rpt_edge == 2'b00) begin
a = 1'b1; b=1'b1; end

if(rpt_edge == 2'b01) begin
a = 1'b1; b=1'b0; end

if(rpt_edge == 2'b10) begin
a = 1'b0; b=1'b1; end

for(i=(repetition-1); i<(repetition+3); i++)
begin
  repeat(1) @(posedge clk);
  a <= ~a;
  repeat(start_wait) @(posedge clk);
  b <= ~b;

  // consecutive repeat condition

    repeat((i)) @(posedge clk);
    b <= ~b;
    stop_wait <= $random() % 4;
    repeat(stop_wait[0]) @(posedge clk);
    a <= ~a;
  end
```

Figure 7-10 shows a sample waveform for a "repeat after" condition. Figure 7-11 shows a sample waveform for a "repeat until" condition.

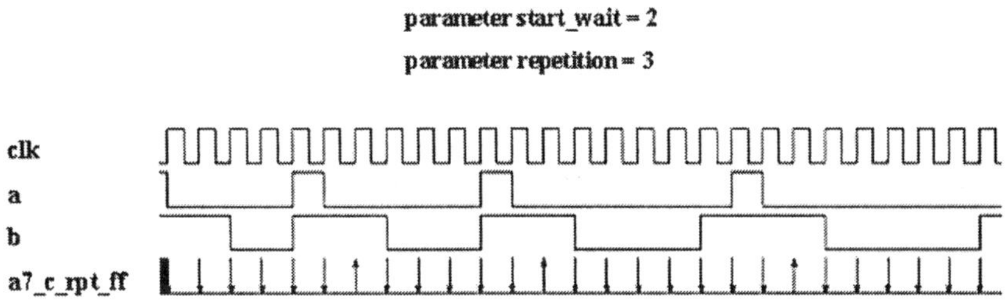

Figure 7-10. Waveform for "repeat after" condition

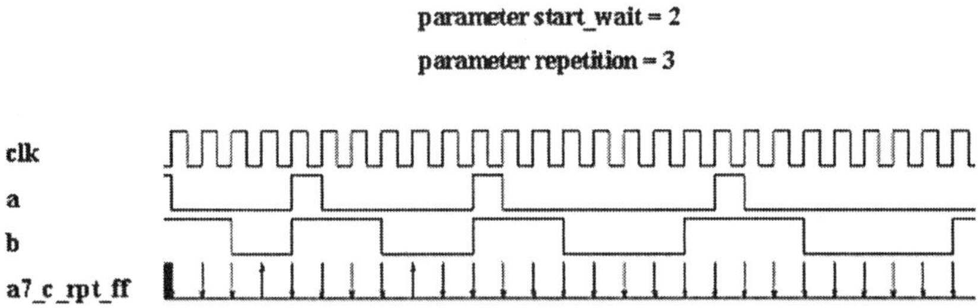

Figure 7-11. Waveform for "repeat until" condition

A few possible relationships between 2 signals were discussed so for. The SVA constructs are abundant and there is a big list of possible relationships between 2 signals. We are not trying to produce a solution for all of those cases. What we have is a small part of the solution. The solution becomes more difficult as the number of signals involved increase.

7.2.7 Environment for ATB involving two signals

In the last few sections, we saw how an exhaustive set of stimulus can be generated to test different relationships between two signals. Several sample Verilog test codes that can satisfy different relationships were shown. In this section, we put the pieces together to create a single configurable testing structure, as shown in Figure 7-12.

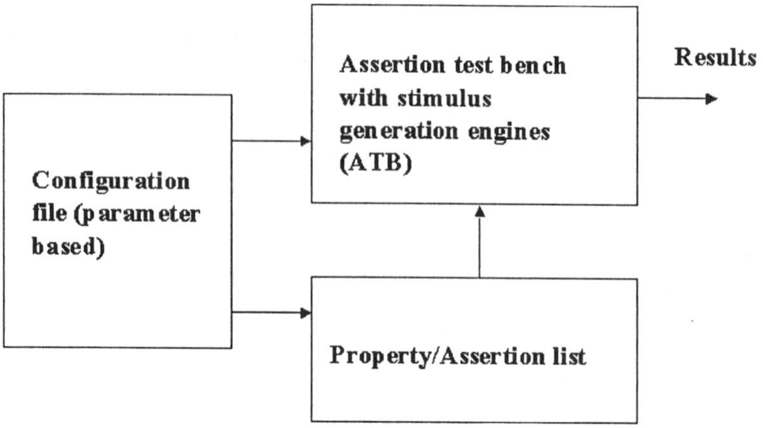

Figure 7-12. ATB Environment

There are three parts in the testing structure:

1. A parameter based configuration file wherein the user can specify the relationship he wishes to test.
2. A SVA listing file containing both the property definitions and the code that will selectively assert properties, based on the parameter configuration.
3. The top-level Verilog test code containing the various stimulus generation blocks discussed in the previous sections. Based on the parameter configuration, the relevant stimulus generation block will be executed to thoroughly verify the current property under test. The parameter definitions are shown in Table 7-1.

Table 7-1. Parameter definitions

Parameter	Functionality
parameter sig_edge = 0;	Defines if signals involved are edge sensitive, 0 indicates no, 1 indicates yes
parameter sig1_edge = 1;	Defines the edge of the leading signal, 1 means rising edge and 0 means falling edge (used only for logical relationship)
parameter logic_op = 0;	Defines if the assertions involve logical relationship, 0 indicates no, 1indicates yes
parameter timing = 1;	Defines if the assertions involve temporal relationship, 0 indicates no, 1 indicates yes
parameter min_time = 2;	Defines timing information, maximum time should be

Parameter	Functionality
parameter max_time = 2;	greater than the minimum time, minimum time cannot be zero, maximum time should be bounded
parameter timing_level = 2'b10;	Defines levels/edges of the signals, "10" means HL for level and RF for edge
parameter o_l_implication = 0;	Parameter to indicate overlapping implication with level sensitive signals, 0 means no, 1 is yes
parameter o_e_implication = 0;	Parameter to indicate overlapping implication with edge sensitive signals
parameter non_o_implication = 1;	Parameter to indicate non-overlapping implication
parameter rpt_me = 0;	Parameter to indicate repetitions are involved
parameter rpt_edge = 2'b00;	Parameter defining the levels/edges of signals
parameter start_wait = 2;	Paramter to define the wait period before repetition in "repeat after"
parameter repetition = 3;	Parameter to define number of repetitions. Repetition value has to be greater than 1
parameter c_rpt = 0;	Parameter indicating the repeat after test
parameter c_rpt_until = 0;	Parameter indicating the repeat until test

By setting the right parameters, a user can generate stimulus for a specific relationship.

- If the parameter "logic_op" is set to 1 and the other parameters are set to 0, then the ATB will generate stimulus for "logical relationship" between two level sensitive signals.

- If the parameter "timing" is set to 1 and all other parameters are set to 0, then the ATB will generate stimulus for "timing relationship" between two signals. A user can specify whether he wants a fixed time or a variable time relation by setting the values of the parameters "min_time" and "max_time." If the user sets the "sig_edge" parameter to 1, then the signals will be treated as edge sensitive.

A sample SVA listing file used for the verification of SVA involving two signals is shown below.

```
module sig_sva (a, b, clk);

// include the parameter definitions
```

```
`include "config.v"

input logic a, b, clk;

// List definitions of all properties under test

// code to selectively include assertions

always@(posedge clk)
begin

// logical relationship between two level
// sensitive signals

if(logic_op == 1 && timing == 0 && sig_edge == 0)
begin
a_1_hh : assert property(p_1_hh);
a_1_hl : assert property(p_1_hl);
a_1_lh : assert property(p_1_lh);
a_1_ll : assert property(p_1_ll);
end

// logical relationship between two edge
// sensitive signals FF,FR

if(logic_op == 1 && timing == 0 && sig_edge == 1
&& sig1_edge == 0)
begin
a2_ff: assert property(p2_ff);
a2_fr: assert property(p2_fr);
end

// logical relationship between 2 edge sensitive
// signals RF,RR

if(logic_op == 1 && timing == 0 && sig_edge == 1
&& sig1_edge == 1)
begin
a2_rf: assert property(p2_rf);
a2_rf: assert property(p2_rr);
end

// timing relationship between 2 level sensitive
```

```
// signals

if(logic_op == 0 && timing == 1 && sig_edge == 0
&& non_o_implication == 0)
begin
if(min_time == max_time)
begin
if(timing_level == 2'b11)
  a3_hh: assert property(p3_hh);
if(timing_level == 2'b10)
  a3_hl: assert property(p3_hl);
if(timing_level == 2'b01)
  a3_lh: assert property(p3_lh);
if(timing_level == 2'b00)
  a3_ll: assert property(p3_ll);
end
if(min_time != max_time)
begin
if(timing_level == 2'b11)
  a3_w1_hh: assert property(p3_w1_hh);
if(timing_level == 2'b10)
  a3_w2_hl: assert property(p3_w2_hl);
if(timing_level == 2'b01)
  a3_w3_lh: assert property(p3_w3_lh);
if(timing_level == 2'b00)
  a3_w4_ll: assert property(p3_w4_ll);
end
end

// logical relationship between 2 level sensitive
// signals with overlapping implication

if((logic_op == 1 || o_l_implication == 1) &&
timing == 0 && sig_edge == 0)
begin
a4_oli_hh: assert property(p4_oli_hh);
a4_oli_hl: assert property(p4_oli_hl);
a4_oli_lh: assert property(p4_oli_lh);
a4_oli_ll: assert property(p4_oli_ll);
end

// logical relationship between 2 edge sensitive
// signals with overlapping implication FF, FR
```

```
if((logic_op == 1 || o_e_implication == 1) &&
timing == 0 && sig_edge == 1 && sig1_edge == 0)
begin
a4_oei_ll: assert property(p4_oei_ll);
a4_oei_lh: assert property(p4_oei_lh);
end

// logical relationship between 2 edge sensitive
// signals with overlapping implication RF, RR

if((logic_op == 1 || o_e_implication == 1) &&
timing == 0 && sig_edge == 1 && sig1_edge == 1)
begin
a4_oei_hl: assert property(p4_oei_hl);
a4_oei_hh: assert property(p4_oei_hh);
end

if(logic_op == 0 && timing == 1 && sig_edge == 1
&& non_o_implication == 0)
begin
if(min_time == max_time)
begin
a4_f_rr: assert property(p4_f_rr);
a4_f_ff: assert property(p4_f_ff);
a4_f_rf: assert property(p4_f_rf);
a4_f_fr: assert property(p4_f_fr);
end
if(min_time != max_time)
begin
a4_w_rr: assert property(p4_w_rr);
a4_w_ff: assert property(p4_w_ff);
a4_w_rf: assert property(p4_w_rf);
a4_w_fr: assert property(p4_w_fr);
end
end

// timing relation between 2 level sensitive
// signals with non-overlapping implication

if(logic_op == 0 && timing == 1 && sig_edge == 0
&& non_o_implication == 1)
begin
if(min_time == max_time)
```

```
begin
if(timing_level == 2'b11)
 a5_f_hh: assert property(p5_f_hh);
if(timing_level == 2'b10)
 a5_f_hl: assert property(p5_f_hl);
if(timing_level == 2'b01)
 a5_f_lh: assert property(p5_f_lh);
if(timing_level == 2'b00)
 a5_f_ll: assert property(p5_f_ll);
end
if(min_time != max_time)
begin
if(timing_level == 2'b11)
 a5_w_hh: assert property(p5_w_hh);
if(timing_level == 2'b10)
 a5_w_hl: assert property(p5_w_hl);
if(timing_level == 2'b01)
 a5_w_lh: assert property(p5_w_lh);
if(timing_level == 2'b00)
 a5_w_ll: assert property(p5_w_ll);
end
end

// timing relation between 2 edge sensitive
// signals with non-overlapping implication

if(logic_op == 0 && timing == 1 && sig_edge == 1
&& non_o_implication == 1)
begin

if(min_time == max_time)
begin
a6_f_rr: assert property(p6_f_rr);
a6_f_ff: assert property(p6_f_ff);
a6_f_rf: assert property(p6_f_rf);
a6_f_fr: assert property(p6_f_fr);
end
if(min_time != max_time)
begin
a6_w_rr: assert property(p6_w_rr);
a6_w_ff: assert property(p6_w_ff);
a6_w_rf: assert property(p6_w_rf);
a6_w_fr: assert property(p6_w_fr);
```

```
end
end

// repetition relationship

if(rpt_me == 1 && c_rpt == 1 && rpt_edge ==
2'b11)
begin
a7_c_rpt_rh: assert property(p7_c_rpt_rh);
a7_cu_rpt_rh: assert property(p7_cu_rpt_rh);
end

if(rpt_me == 1 && c_rpt == 1 && rpt_edge ==
2'b10)
begin
a7_c_rpt_rl: assert property(p7_c_rpt_rl);
a7_cu_rpt_rl: assert property(p7_cu_rpt_rl);
end

if(rpt_me == 1 && c_rpt == 1 && rpt_edge ==
2'b01)
begin
a7_c_rpt_fh: assert property(p7_c_rpt_fh);
a7_cu_rpt_fh: assert property(p7_cu_rpt_fh);
end

if(rpt_me == 1 && c_rpt == 1 && rpt_edge ==
2'b00)
begin
a7_c_rpt_fl: assert property(p7_c_rpt_fl);
a7_cu_rpt_fl: assert property(p7_cu_rpt_fl);
end
end

// Config Parameters illegal values. If
// logical_op is asserted then timing cannot be
// asserted

property config_check1;
  @(posedge clk)
       (logic_op == 1) |->
                         (timing == 0);
endproperty
```

```
// only one of the implication operators can be
// asserted at any time

property config_check2;
  @(posedge clk)
      $onehot0({o_l_implication,
      o_e_implication, non_o_implication});
endproperty

// min_time should be atleast 1

property config_check3;
  @(posedge clk)
    (timing == 1) |-> (min_time >= 1);
endproperty

// repetition should be greater that one

property config_check4;
  @(posedge clk)
      ((c_rpt == 1) && (rpt_me == 1)) |->
                  (repetition > 1);
endproperty

a_check1: assert property(config_check1);
a_check2: assert property(config_check2);
a_check3: assert property(config_check3);
a_check4: assert property(config_check4);

 endmodule
```

The ATB gets executed based on the parameter configuration. A sample ATB used for the verification of SVA involving two signals is shown below.

```
module sig_sva_tb;

logic a,b;
logic clk;
logic [1:0] rpt_wait;
logic [1:0] stop_wait;

`include  "config.v"

integer i,j;
```

```
logic [1:0] logical_op_reg;

initial
begin
clk = 1'b0; a=1'b0; b=1'b0;
logical_op_reg = 2'b00;

//*********************************************
// case 1
// logical operation, overlapping implication
// level sensitive signals
//*********************************************

if((logic_op == 1 || o_l_implication == 1) &&
timing == 0 && sig_edge == 0)
begin
for(i=0; i<4; i++)
begin
  a <= logical_op_reg[0];
  b <= logical_op_reg[1];
  repeat(1) @(posedge clk);
  logical_op_reg++;
end
end

//*********************************************
// case 2
// logical operation, overlapping implication
// edge sensitive signals
//*********************************************

if((logic_op == 1 || o_e_implication == 1) &&
timing == 0 && sig_edge == 1)
begin

if(sig1_edge == 0) // ff, fr
begin
for(i=0; i<8; i++)
begin
  a <= logical_op_reg[0];
  b <= logical_op_reg[1];
  repeat(1) @(posedge clk);
  logical_op_reg++;
```

```
end
end

if(sig1_edge == 1) // rr, rf
begin
for(i=0; i<8; i++)
begin
 a <= !logical_op_reg[0];
 b <= !logical_op_reg[1];
 repeat(1) @(posedge clk);
 logical_op_reg++;
end
end

end

//*************************************************
// case 3
// timing relation between 2 signals
//*************************************************

if(logic_op == 0 && timing == 1)
begin

if(timing_level == 2'b11) begin
a = 1'b0; b=1'b0;
end

if(timing_level== 2'b00) begin
a = 1'b1; b=1'b1;
end

if(timing_level == 2'b01) begin
a = 1'b1; b=1'b0;
end

if(timing_level == 2'b10) begin
a = 1'b0; b=1'b1;
end
for(i=(min_time-1); i<(max_time+3); i++)
begin
 repeat(1) @(posedge clk);
 a <= ~a;
```

```
    if(i == 0)
    begin
      b <= ~b;
      repeat(1) @(posedge clk);
      a <= ~a; b <= ~b;
    end
    else
    begin
      repeat(1) @(posedge clk);
      a<= ~a;
      repeat((i-1)) @(posedge clk);
      b<= ~b;
      repeat(1) @(posedge clk);
      b<= ~b;
    end
  end
  end

  //***************************************************
  // case 4
  // repetitions
  //***************************************************

  if(rpt_me == 1)
  begin

  if(rpt_edge == 2'b11) begin
  a = 1'b0; b=1'b0;
  end

  if(rpt_edge == 2'b00) begin
  a = 1'b1; b=1'b1;
  end

  if(rpt_edge == 2'b01) begin
  a = 1'b1; b=1'b0;
  end

  if(rpt_edge == 2'b10) begin
  a = 1'b0; b=1'b1;
  end

  if(c_rpt == 1)
```

```
begin
for(i=(repetition-1); i<(repetition+3); i++)
begin
  repeat(1) @(posedge clk);
  a <= ~a;
  repeat(start_wait) @(posedge clk);
  b <= ~b;

  // consecutive repeat condition

  repeat((i)) @(posedge clk);
  b <= ~b;
  stop_wait <= $random() % 4;
  repeat(stop_wait[0]) @(posedge clk);
  a <= ~a;
end
end
end

repeat(2) @(posedge clk);
$finish();
end

initial
forever clk = #25 ~clk;

endmodule

bind sig_sva_tb sig_sva i1 (a, b, clk);
```

Note that the configuration file is included into the ATB and the SVA listing file is bound to the ATB module.

7.3 ATB example for a PCI Checker

In this section, we take a complex SVA checker and show how to verify its functionality based on the concepts discussed in Section 7.2. Following is a checker discussed in Chapter 6.

"Target latency for the completion of the first data phase is 16 cycles from the assertion of the signal `framen`"

```
sequence s_tchk9a;
@(posedge clk)
(!irdyn && !trdyn);
endsequence

sequence s_tchk9b;
@(posedge clk)
(!irdyn && !stopn);
endsequence

sequence s_tchk9_fast;
@(posedge clk)
$fell(framen) ##1 $fell(devseln);
endsequence
property p_tchk9_fast;
@(posedge clk)
s_tchk9_fast |->
(!framen && !devseln) throughout
(##[1:15] (s_tchk9a.ended || s_tchk9b.ended));
endproperty

a_tchk9_fast: assert property(p_tchk9_fast);
c_tchk9_fast: cover property(p_tchk9_fast);
```

The property p_tchk9_fast involves complex logical relationship and temporal relationship. The property becomes active when a valid start condition happens (s_tchk9_fast). Once the antecedent matches, the property can succeed in two different ways. Either s_tchk9a or s_tchk9b should match within 1 to 15 cycles, assuming the two signals that helped the antecedent match remain asserted throughout the evaluation.

To verify this check **exhaustively**, the following should be done:

1. All possible logical relationships between the signals "irdyn," "trdyn," "devseln" and "stopn" should be tested.
2. All possible temporal relationships between the antecedent and the consequent should be tested.

Based on the theory from Section 7.2, the possible logical combination for the four signals is 16, as shown in Table 7-2. Upon a successful match of the antecedent, any one of the success condition shown in Table 7-2 should occur within 1 to 15 clock cycles. By looping the checker from "(min-1)" to "(max+1)," which is 0 to 16, one can simulate all possible successes for all possible delay conditions and at least one error condition. While looping in a specific delay value, all 16 possible logical relationships should be executed. Hence, there are 16 possible time slots and 16 possible logical conditions, leading to 256 possible scenarios. The check will pass on three conditions in each of the delay values starting from 1 to 15 and hence, there should be 45 real successes for this check. In other words, this check can succeed in 45 different conditions. This metric can be cross checked by writing cover statements for the property under validation.

Table 7-2. Logical conditions for PCI check

irdyn	trdyn	devseln	stopn	Status
0	0	0	0	Success (s_tchk9a && s_tchk9b)
0	0	0	1	Success (s_tchk9a)
0	0	1	0	Failure
0	0	1	1	Failure
0	1	0	0	Success (s_tchk9b)
0	1	0	1	Failure
0	1	1	0	Failure
0	1	1	1	Failure
1	0	0	0	Failure
1	0	0	1	Failure
1	0	1	0	Failure
1	0	1	1	Failure
1	1	0	0	Failure
1	1	0	1	Failure
1	1	1	0	Failure
1	1	1	1	Failure

A sample Verilog code that generates stimulus to satisfy these 256 different scenarios is shown below.

```
module ctc_complex;

logic irdyn, trdyn, devseln, stopn, framen;
integer i, j;
logic clk;
logic [3:0] test_expr;

assign irdyn = test_expr[3];
assign trdyn = test_expr[2];
assign devseln = test_expr[1];
assign stopn = test_expr[0];
```

```
initial begin
clk = 1'b0; test_expr = 4'd15;
repeat(2) @(posedge clk);

for(i=1; i<17; i++) // timing loop
begin
for(j=0; j<16; j++) // logical loop
begin
  framen = 1'b0;
  repeat(1) @(posedge clk);
  test_expr = 4'b1101;
  repeat(i) @(posedge clk);
  test_expr = j;
  repeat(1) @(posedge clk);
  framen = 1'b1;
  repeat(1) @(posedge clk);
  test_expr = 4'b1111;
  repeat(1) @(posedge clk);
end
end

repeat(2) @(posedge clk);
$finish;
end

initial forever clk = #25 ~clk;

endmodule
```

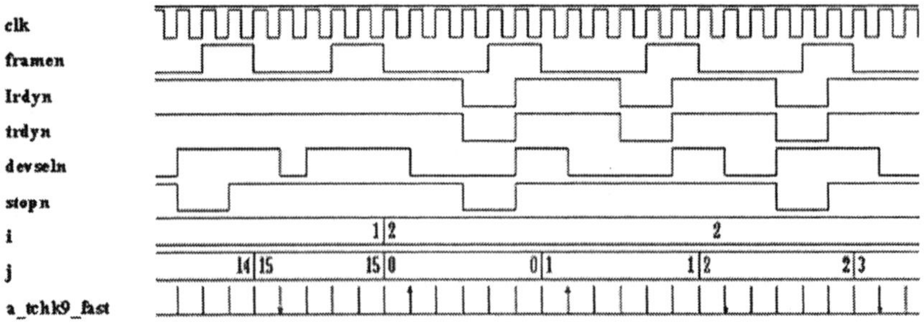

Figure 7-13. PCI Checker verification

Note that the timing constraint is used as the outer loop and the logical constraint as the inner loop. The simulation of this test code on the checker should produce 211 failures and 45 successes. This guarantees that the checker responds correctly for all possible input conditions. A sample waveform from this test is shown in Figure 7-13.

7.4 Summary for checking the checker

- The possible number of relationships between two signals can grow exponentially within the SVA domain.
- By exploring just three of these relationships (logic, timing and repetition) between 2 signals, 56 possible assertion statements were written. As the number of signals involved in an SVA definition increases, the possible assertion statements will increase exponentially.
- It is critical to have an automated way to validate these assertions. With basic Verilog language we were able to create some very effective stimulus generation schemes that tested the assertions thoroughly.
- The same stimulus generation methodology was applied on a real life PCI checker to verify its correctness.
- As the assertions get more complex, the advanced features of constrained random testbenches can be used effectively to check these assertions. Self checking mechanisms can be used to analyze the results of the assertion validation.
- While there is no automated way of checking the checker yet, a user can still verify them like any other design module using testbenches.
- Without a "checking the checker" methodology, a user will not know if the design is working or failing, or if the checker was written incorrectly.
- This process demands some time investment in the beginning of verification process, but can be a huge time saver in the latter part of verification.

References

[1] SystemVerilog 3.1a_LRM.pdf from Accellera

[2] PCI Local Bus Specification Revision 2.1s

[3] PCI System Architecture, Third Edition – Tom Shanley and Don Anderson

[4] System Verilog for Design – Stuart Sutherland, Simon Davidmann and Peter Flake

[5] Circuits and Systems, 1999. 42nd Midwest Symposium on Volume 1, Aug. 1999, A New Gate Image Decoder; Algorithm, Design and Implementation – Reza Hashemian and Srikanth Vijayaraghavan

[6] Synopsys Designware Memory Controller DW_memctl MacroCell Application Notes

[7] Synopsys PCI/PCI-X Flexmodel User's Manual (PCI 2.3, PCI-X1.0, PCI-X2.0)

[8] Intel Strata Flash Memory(J3) – Datasheet

[9] Samsumg CMOS SRAM – Datasheet

[10] Micron SDRAM – Datasheet

[11] Elpida 512bits DDR SDRAM – Datasheet

CD-ROM DISCLAIMER

INDEX

Functions

Symbols

A

B

C

D

E

F